Analog Circuits and Signal Processing

Series editors

Mohammed Ismail, Dublin, USA
Mohamad Sawan, Montreal, Canada

More information about this series at http://www.springer.com/series/7381

João Carlos Ferreira de Almeida Casaleiro
Luís Augusto Bica Gomes Oliveira
Igor M. Filanovsky

Quadrature RC–Oscillators

The van der Pol Approach

Springer

João Carlos Ferreira de Almeida Casaleiro
Instituto Politécnico de Lisboa
ISEL-Inst Sup de Engenharia de Lisboa
Lisboa, Portugal

Luís Augusto Bica Gomes Oliveira
Universidade Nova de Lisboa
Caparica, Portugal

Igor M. Filanovsky
University of Alberta
Edmonton, AB, Canada

ISSN 1872-082X ISSN 2197-1854 (electronic)
Analog Circuits and Signal Processing
ISBN 978-3-030-00739-3 ISBN 978-3-030-00740-9 (eBook)
https://doi.org/10.1007/978-3-030-00740-9

Library of Congress Control Number: 2018955496

This Springer imprint is published by the registered company Springer Nature Switzerland AG
The registered company address is: Gewerbestrasse 11, 6330 Cham, Switzerland

To the authors' families

Acknowledgements

The authors thank colleagues at Instituto Superior de Engenharia de Lisboa, Universidade Nova de Lisboa, CTS-UNINOVA and INESC-ID Lisboa for their support and contributions to the work presented in this book. A special thanks to Professor Manuel M. Silva for the helpful discussions, his illuminating suggestions to clarify the text and meticulous reviews of this book.

This work was supported by the Portuguese Foundation for Science and Technology (FCT/MCTES) (CTS multi-annual funding) through PIDDAC programme funds and under projects DISRUPTIVE (EXCL/EEI-ELC/0261/2012), PEST (PESTOEEEI/UI0066/2014), and foRESTER (PCIF/SSI/0102/2017).

The authors acknowledge the support given by the following institutions:

- CTS-UNINOVA (Center of Technology and Systems of Universidade Nova de Lisboa)
- INESC-ID Lisboa (Instituto de Engenharia de Sistemas e Computadores - Investigação e Desenvolvimento em Lisboa)

for providing access to their integrated circuit design and laboratory facilities.

Contents

Acronyms

AAC	Automatic amplitude control
CMOS	Complementary metal-oxide-semiconductor
FoM	Figure-of-merit
ILFD	Injection-locked frequency divider
IoT	Internet-of-things
IRR	Image rejection ratio
ISM	Industrial, scientific and medical
KCL	Kirchhoff's current law
KVL	Kirchhoff's voltage law
OFDM	Orthogonal frequency division multiplexing
QAM	Quadrature amplitude modulation
QO	Quadrature oscillator
QPSK	Quadrature phase-shift keying
QVCO	Quadrature voltage-controlled oscillator
RF	Radio frequency
SNR	Signal-to-noise ratio
THD	Total harmonic distortion
VDP	Van der Pol
VDPO	Van der Pol oscillator
WSN	Wireless sensor networks

Chapter 1
Introduction

1.1 Background and Motivation

1.1.1 Quadrature Oscillators

The objective of an oscillator circuit is to generate a periodic signal. Radio communication frequency circuits require sinusoidal oscillators with well-controlled amplitude, frequency and phase. Digital circuits and analogue-to-digital converters require square-wave signals, usually referred to as clock signals. These are generated by relaxation oscillators (also called first-order oscillators or multivibrators), a topic that is outside the scope of this book, which is focused on sinusoidal oscillators. Modern radio frequency (RF) receiver architectures, like the low-IF receiver shown in Fig. 1.1, require two sinusoidal oscillators, L_{O1} and L_{O2}, with low phase noise and signals which are accurately in quadrature to reject the image band [1, 2]. The sensitivity of the receiver is limited by the image rejection ratio (IRR), which, in turn, is limited by the circuit mismatches and quadrature error. This makes the quadrature oscillator (QO) a key block in the receivers.

In recent years, a significant research effort has been invested in the study of oscillators with accurate quadrature outputs, with less than 1° of phase error. The demand for low-power QOs that generate accurate quadrature signals has been growing with the widespread adoption of digital communications, in systems that use quadrature amplitude modulation (QAM) and quadrature phase-shift keying (QPSK). Examples of standards are Zigbee (IEEE 802.15.4) and Bluetooth (IEEE 802.15.1), used in a multitude of applications, such as wireless sensor networks (WSN), home automation, healthcare, smart energy power supplies and many others. Moreover, in modern receivers the cost and size reduction are important requirements. The minimization of external components is required to reduce the equipment cost. However, full integration poses several challenges. For instance, the low-IF receiver requires image cancellation. This can be obtained if quadrature

J. C. F. d. A. Casaleiro et al., *Quadrature RC–Oscillators*, Analog Circuits and Signal Processing, https://doi.org/10.1007/978-3-030-00740-9_1

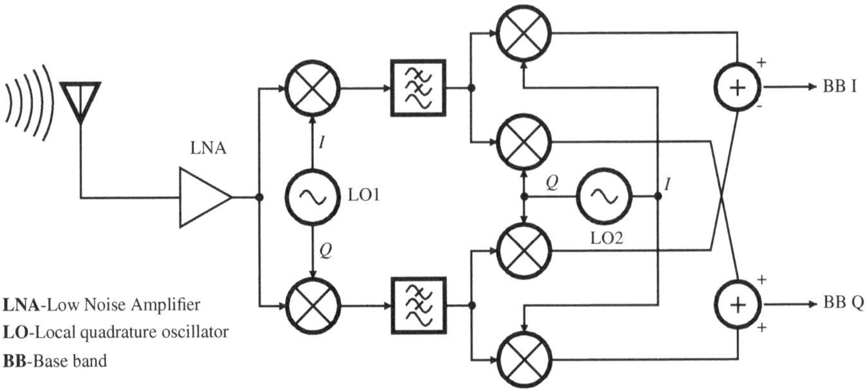

Fig. 1.1 Low-IF receiver front-end block diagram

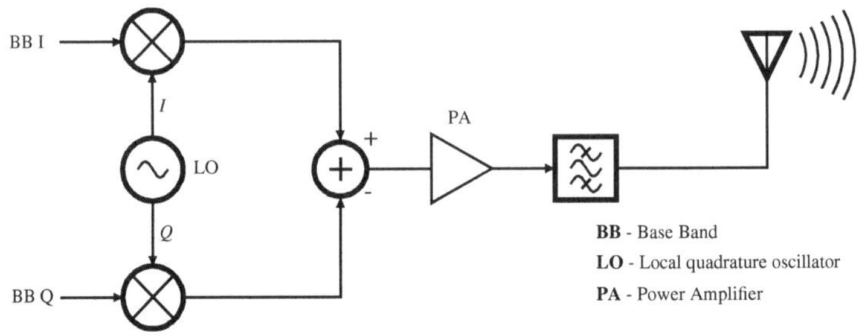

Fig. 1.2 Direct up-conversion transmitter block diagram

signals are available, thus avoiding image-rejection filters that require a large die area [1, 3].

Modern RF transmitter architectures, with direct up-conversion (an example is shown in Fig. 1.2), using spectrum-efficient modulations, such as orthogonal frequency division multiplexing (OFDM), also require QO with accurate quadrature signals. The quadrature error can limit the achievable signal-to-noise ratio (SNR) and the size of the supported constellation and data rates.

1.1.2 Quadrature Signal Generation

Several methods to generate quadrature signals are described in the literature. In this section, we review first two open-loop approaches. The first of them is the use of *RC–CR* networks and the other is the use of polyphase filters. These are passive networks that need large die area and high input power (to overcome the attenuation imposed by the filtering) to achieve acceptable quadrature inaccuracies (below 1°) [1, 3]. Afterwards, the frequency division method is reviewed.

Fig. 1.3 *RC–CR* circuit

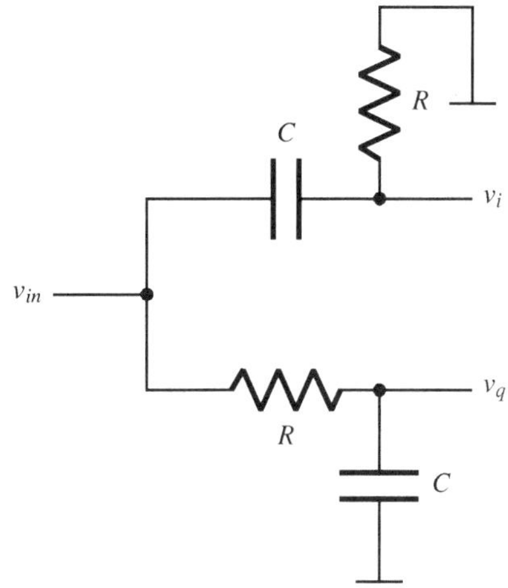

1.1.2.1 *RC–CR* Network

This method splits the input signal v_{in} into two by passing it through the RC and CR branches, as shown in Fig. 1.3. The CR branch is a high-pass filter that shifts the output signal phase by $+45°$ and the RC branch is a low-pass filter that shifts the output signal phase by $-45°$ with respect to the input signal, at the pole frequency $\omega_p = 1/(RC)$. Although the phase shift varies in each branch with the frequency, the phase difference between the signals of the two branches is always $90°$. However, the branches' attenuations are equal only at the pole frequency [2, 3].

In theory, this is not a problem because one can design the network so that the pole frequency is equal to the oscillator frequency. In practice, however, due to the temperature and process variations, one cannot guarantee either the absolute value of the network components or a perfect match between them. Hence, in practical circuits there are amplitude and quadrature errors. To minimize the errors, more stages can be added to the network, as shown in Fig. 1.4. The *RC–CR* network of Fig. 1.4 is known as a polyphase filter. By adding more stages, the errors decrease and the operating bandwidth increases, but the signal loss increases considerably.

1.1.2.2 Frequency Division

Frequency division provides wideband quadrature generation. However, the divider-by-two method requires twice the nominal frequency of operation. This increases the power requirements, especially at high frequencies [2]. The divider consists of two latches connected in a master/slave configuration, as shown in Fig. 1.5a. A square-wave input signal with 50% duty cycle is used as clock signal to generate two quadrature output signals, as shown in Fig. 1.5b.

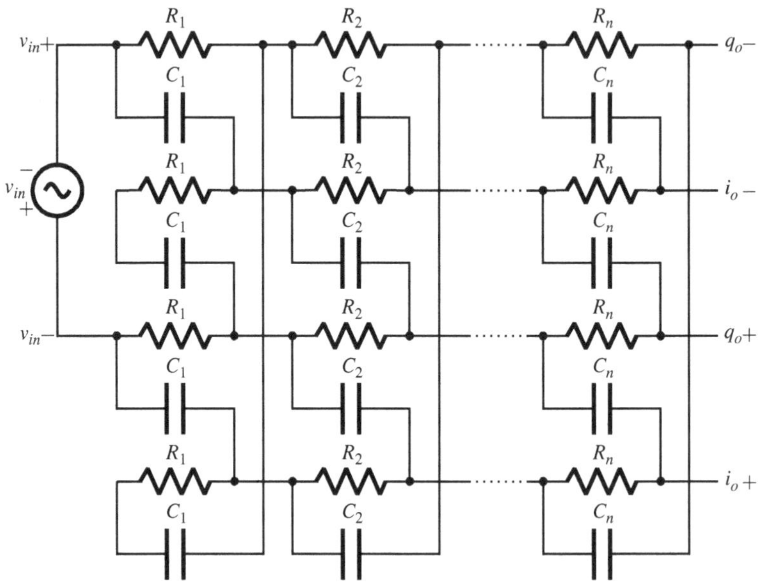

Fig. 1.4 Passive *RC* polyphase filter with four phases and *n* stages

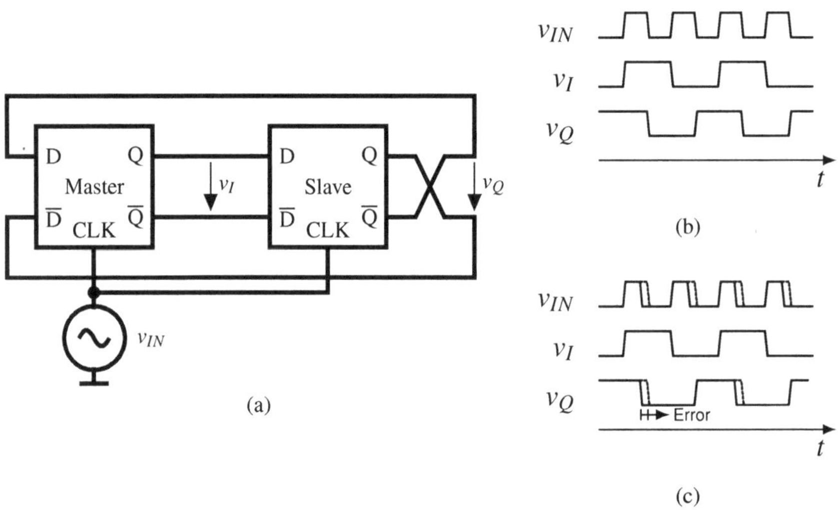

Fig. 1.5 Digital frequency divider-by-two: (**a**) circuit, (**b**) waveforms and (**c**) waveforms with phase error

If the input signal does not have 50% duty cycle, then the output signals have a quadrature error, as shown in Fig. 1.5c. This problem can be solved by using a divider-by-four, but, in this case, the frequency of the input signal must have four times the desired output signal frequency.

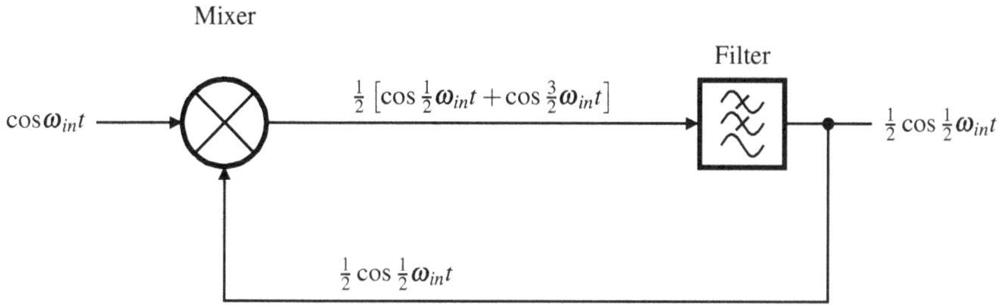

Fig. 1.6 Regenerative divider-by-two

The divider-by-two based on latches is, in addition, inadequate to produce quadrature sinusoidal signals because the outputs are square waves. Hence, it requires additional filtering that needs a large chip area to cope with the components' mismatches.

For sinusoidal outputs, dynamic frequency dividers, such as the injection-locked frequency divider (ILFD) [4, 5] or the regenerative divider [6, 7], are more adequate.

The regenerative divider, also known as Miller divider, consists of a mixer and a low-pass filter connected in a feedback structure, as shown in Fig. 1.6. If we assume that the input signal is sinusoidal, with frequency ω_{in}, and a feedback signal is also sinusoidal, with half frequency $\omega_{\text{in}}/2$, then the signal after mixing should have the frequency components $\omega_{\text{in}}/2$ and $3\omega_{\text{in}}/2$. The higher frequency is attenuated by the low-pass filter. Hence, the harmonic purity of the output signal depends on the attenuation imposed by the low-pass filter.

In general, these open-loop methods have worse performance than the closed-loop ones, investigated in this book. Also, open-loop methods do not allow the compensation of the mismatches.

1.1.2.3 Coupled Oscillators

The closed-loop approaches include coupled oscillators and ring oscillators. The best QOs are based on two coupled LC-oscillators: they have the lowest phase noise and phase error [8]. Recently, it was shown that they can achieve perfect quadrature [9]. However, coupled LC-oscillators require two inductors, which, depending on the frequency, can occupy a large die area. Moreover, inductors do not scale down with the technology, and designing inductors with acceptable quality factor ($Q > 5$) requires the use of thick top metal layers, which increases the chip cost [10]. Inductorless QOs, like the two-integrator oscillator or coupled RC-oscillators, are viable alternatives to avoid the use of inductors. However, in comparison with coupled LC-oscillators both coupled and ring RC-oscillators have poorer phase-noise performance [8]. For industrial, scientific and medical (ISM) band, the phase noise of inductorless oscillators may satisfy the requirements. For instance, the phase-noise specification for 2.4 GHz ISM band at the offset of 1 MHz from the carrier is -110 dBc/Hz for Bluetooth and -88 dBc/Hz for Zigbee; these values are within the performance capability of inductorless oscillators [11].

The analysis of sinusoidal oscillators using the linear positive feedback model is usually sufficient for deriving the oscillation frequency. However, due to the circuit linearization, as will be shown below, the amplitude limit mechanism is lost, since it is dependent on the circuit nonlinearities. A large-signal analysis can overcome this limitation, but leads to long and complicated equations that do not help the designer. In this book, the analysis based on the weak nonlinearity of the transistors' transconductances is presented. It is shown the similarity of these weak nonlinearities to that of the van der Pol oscillator model presented in [12]. The van der Pol oscillator model describes both the amplitude limitation and the frequency selectivity. This approach allows to avoid a large-signal analysis. Moreover, since the van der Pol oscillator model has been extensively studied, it is used in this book for the analysis of coupled oscillators as well.

Coupled oscillators consist of two identical oscillators connected by either an active or a passive network. Several active coupling networks were proposed; they can be grouped into either parallel or series topologies. The parallel topology was first proposed in [13] for LC-oscillators, with the coupling of amplifier transistors in parallel with the oscillators' core. In the series topology, proposed in [14], the transistors are in series with the oscillators' core. A comprehensive comparison of these two topologies for LC-oscillators can be found in [15]. The disadvantage of the parallel topology is the use of two extra gain blocks, which increases the power dissipation [2]. The series topology reuses the current of the oscillator, but the output swing is limited and their application is limited as well, since the trend in future complementary metal-oxide-semiconductor (CMOS) technologies is to lower the supply voltages towards 0.5–0.7 V [16]

With passive coupling, the amplifiers are substituted by passive elements (usually inductors or capacitors). The coupling based on inductors [17] and transformers [18, 19] requires a larger area than active coupling. Capacitive coupling of LC-oscillators has shown specific results [20]: as opposed to traditional active coupling, it does not increase the power consumption. However, the area minimization is still limited by the inductors, and the oscillation frequency is lower [21].

In this book, three quadrature oscillators working in the sinusoidal regime are investigated: RC-oscillator with active coupling, RC-oscillator with capacitive coupling, and the two-integrator oscillator. Of special interest is the study of quadrature RC-oscillator with capacitive coupling [22]. The capacitive coupling is noiseless and requires a small area. Since the coupling capacitors do not add noise, we expect a 3-dB phase-noise improvement (due to coupling), and, with a marginal increase of the power, a figure-of-merit (FoM) comparable to that of the best state-of-the-art RC-oscillators is achieved. Contrarily to what might be expected, with the increase of the coupling capacitances (higher coupling strength) the oscillation frequency increases [22]. We present the theory to explain this behaviour and derive the equations for the frequency, phase error and amplitude mismatch; the equations are validated by simulation. The theory shows that the phase error is proportional to the amplitude mismatch, indicating that an automatic phase-error minimization based on the amplitude mismatch is possible. We also study bimodal oscillations and phase ambiguity, for this coupling topology, comparing it with the existing alternatives [23]. To validate the theory, a 2.4-GHz quadrature voltage-controlled oscillator (QVCO) based on two RC-oscillators with capacitive coupling was fabricated, in a standard 0.13 µm CMOS process.

1.2 The Book Organization

This book is organized into eight chapters. In the second chapter, an overview of sinusoidal oscillator models is presented: we describe the positive-feedback and the negative-resistance models. Afterwards, a survey of the automatic amplitude control methods is presented. We focus mainly on the method that uses the intrinsic nonlinearities of the oscillator to limit the amplitude. To conclude the single sinusoidal oscillator modeling, we describe the frequency selectivity and introduce the concept of the oscillator's quality factor.

In Chap. 3, an overview of the van der Pol oscillator (VDPO) model is presented. The weak nonlinearity of the transistors' transconductances is related to the van der Pol oscillator model. The VDPO is used as an example and is analysed using two models presented in the previous chapter.

In Chap. 4, we analyse oscillators driven by an external periodic signal with injecting a current as locking signal. Both the parallel and series topologies are studied, and their locking range is derived. We use the VDPO as a base oscillator for the analysis because its nonlinearities are similar to the nonlinearities of the inductorless oscillators studied in the following chapters.

In Chap. 5, we present the analysis of the actively cross-coupled RC-oscillator, which is a QO that consists of two RC-oscillators coupled by transconductance amplifiers. First, we derive the single RC-oscillator equations that show that a single RC-oscillator can be modeled by the series VDPO. Afterwards, we analyse the quadrature oscillator and derive the frequency, amplitude- and phase-error equations. A stability analysis of the equilibrium point is presented. The theoretical results are validated by simulation.

In Chap. 6, we study the capacitive coupling RC-oscillator regarding the frequency, amplitude and phase error. We investigate the relation between the coupling and the quadrature generation, the impact of the coupling strength on the frequency, amplitude and phase error, and the impact of the mismatches on the amplitude and phase errors. We derive the equations for the frequency, amplitude and phase error as a function of the circuit mismatches. The theoretical results are validated by simulation.

In Chap. 7, we study the two-integrator oscillator working in the quasi-sinusoidal regime. We focus on the investigation of the impact of the circuit mismatches on the frequency, amplitude and phase error. We derive the equation for these key parameters as a function of the circuits' mismatches. The theoretical results are validated by simulation.

Finally, in Chap. 8 we present the conclusions.

1.3 Main Contributions

Several papers describing the main aspects of the proposed material were published in international conferences and journals. To the best of the authors' knowledge, the main contributions of this work are:

- The improvement of the model of the single RC-oscillator (in Chap. 2). We show the relation between the circuit elements and the van der Pol (VDP) parameters.
- A detailed review (in Chap. 3) of the van der Pol oscillator model indicating that VDPO is suitable as a base oscillator for the analysis of modern RC-oscillators in weak and strong nonlinear regime.
- A detailed analysis (in Chap. 4) of the injection locking in van der Pol oscillators.
- A study (in Chap. 5) of the quadrature generation in active coupling RC-oscillators working in the sinusoidal regime. The research is focused on the impact of the mismatches and the coupling strength on the frequency, amplitude and phase errors. The analysis in this chapter differs from other research works: weak coupling strengths are assumed. Other works that analysed this oscillator are assuming a strong coupling (coupling amplifiers work as hard limiters) making the coupling signal a square wave. The theoretical results were validated by simulation.
- A study (in Chap. 6) of the quadrature generation in capacitive coupling RC-oscillators. The research is focused on the impact of the coupling strength on the frequency, amplitude and phase errors [24]. A prototype operating at 2.4 GHz was designed to confirm the theoretical results.
- A study (in Chap. 7), using the VDP approximation, of the two-integrator oscillator in the linear regime. The research is focused on the impact of the coupling strength on the frequency, amplitude and phase errors [25]. The theoretical results were validated by simulation.

References

1. P.-I. Mak, S.-P. U, R. Martins, Transceiver architecture selection: review, state-of-the-art survey and case study. IEEE Circuits Syst. Mag. **7**(2), 6–25 (2007)
2. L. Oliveira, J. Fernandes, I. Filanovsky, C. Verhoeven, M. Silva, *Analysis and Design of Quadrature Oscillators* (Springer, Heidelberg, 2008)
3. B. Razavi, *RF Microelectronics* (Prentice-Hall, Upper Saddle River, 1998)
4. R. Adler, A study of locking phenomena in oscillators. Proc. Inst. Radio Eng. **34**(6), 351–357 (1946)

5. S. Verma, H. Rategh, T. Lee, A unified model for injection-locked frequency dividers. IEEE J. Solid State Circuits **38**(6), 1015–1027 (2003)
6. R. Miller, Fractional-frequency generators utilizing regenerative modulation. Proc. Inst. Radio Eng. **27**(7), 446–457 (1939)
7. J. Lee, B. Razavi, A 40-ghz frequency divider in 0.18-μm CMOS technology. IEEE J. Solid State Circuits **39**(4), 594–601 (2004)
8. L.B. Oliveira, A. Allam, I.M. Filanovsky, J.R. Fernandes, C.J.M. Verhoeven, M.M. Silva, Experimental comparison of phase-noise in cross-coupled RC- and LC-oscillators. Int. J. Circuit Theory Appl. **38**, 681–688 (2010)
9. H. GHonoodi, H.M. Naimi, A phase and amplitude tunable quadrature *LC* oscillator: analysis and design. IEEE Trans. Circuits Syst. I **58**(4), 677–689 (2011)
10. B. Razavi, A study of phase noise in CMOS oscillators. IEEE J. Solid State Circuits **31**(3), 331–343 (1996)
11. N.-J. Oh, S.-G. Lee, J. Ko, A CMOS 868/915 MHz direct conversion. ZigBee single-chip radio. IEEE Commun. Mag. **43**(12), 100–109 (2005)
12. B. van der Pol, The nonlinear theory of electric oscillations. Proc. Inst. Radio Eng. **22**(9), 1051–1086 (1934)
13. A. Rofougaran, J. Rael, M. Rofougaran, A. Abidi, A 900 MHz CMOS LC-oscillator with quadrature outputs, in *IEEE International Solid-State Circuits Conference* (1996), pp. 392–393
14. P. Andreani, A low-phase-noise low-phase-error 1.8 GHz quadrature CMOS VCO, in *IEEE International Solid-State Circuits Conference*, vol. 1 (2002), pp. 290–466
15. P. Andreani, A. Bonfanti, L. Romano, C. Samori, Analysis and design of a 1.8-GHz CMOS LC quadrature VCO. IEEE J. Solid State Circuits **37**(12), 1737–1747 (2002)
16. Radio Frequency and Analog/Mixed-Signal Technologies for Wireless Communications, International Technology Roadmap for Semiconductors, 2007 Edition [Online]
17. A. Willson, Energy circulation quadrature LC-VCO, in *IEEE International Symposium on Circuits and Systems* (IEEE, Kos, 2006), p. 4
18. S. Gierkink, S. Levantino, R. Frye, C. Samori, V. Boccuzzi, A low-phase-noise 5-GHz CMOS quadrature VCO using superharmonic coupling. IEEE J. Solid State Circuits **38**(7), 1148–1154 (2003)
19. W. Li, K.-K.M. Cheng, A CMOS transformer-based current reused SSBM and QVCO for UWB application. IEEE Trans. Microw. Theory Tech. **61**(6), 2395–2401 (2013)
20. C.T. Fu, H.C. Luong, A 0.8-V CMOS quadrature LC VCO using capacitive coupling, in *IEEE Asia Solid-State Circuits Conference (ASSCC'07)* (2007), pp. 436–439
21. L.B. Oliveira, I.M. Filanovsky, A. Allam, J.R. Fernandes, Synchronization of two LC-oscillators using capacitive coupling, in *IEEE International Symposium on Circuits and Systems (ISCAS'08)* (2008), pp. 2322–2325
22. J. Casaleiro, L.B. Oliveira, I. Filanovsky, Low-power and low-area CMOS quadrature RC oscillator with capacitive coupling, in *IEEE International Symposium on Circuits and Systems (ISCAS'12)* (2012), pp. 1488–1491
23. H. Tong, S. Cheng, Y.-C. Lo, A.I. Karsilayan, J. Silva-Martinez, An LC quadrature VCO using capacitive source degeneration coupling to eliminate bi-modal oscillation. IEEE Trans. Circuits Syst. I **59**(9), 1871–1879 (2012)
24. J. Casaleiro, L.B. Oliveira, I.M. Filanovsky, A quadrature RC-oscillator with capacitive coupling. Integr. VLSI J. **52**, 260–271 (2016)
25. J. Casaleiro, L.B. Oliveira, I.M. Filanovsky, Amplitude and quadrature errors of two-integrator oscillator. J. Low Power Electron. **11**, 340–348 (2015)

Chapter 2
Sinusoidal Oscillators

2.1 Introduction

In this chapter, two models of the sinusoidal oscillator are reviewed. First, the linear positive-feedback model and the associated Barkhausen criterion are introduced. Next, the model of the negative-resistance oscillator is considered. For both models, the parallel and series topologies are described. The amplitude control techniques with the main focus on the amplitude limiting by nonlinearities are reviewed. Two implementations of a negative-resistance circuit are presented as examples. The frequency control is briefly discussed.

2.2 Sinusoidal Oscillator Models

Sinusoidal oscillators are usually analysed [1] as linear positive-feedback systems, like the one shown in Fig. 2.1. We will refer to this as the feedback model. The model is suitable for oscillator topologies with a feedback loop, such as the ring and phase-shift oscillators. However, with few exceptions, the feedback model can be used in the analysis of other topologies as well. The model assumes a system composed of a forward network, $H(s)$, a feedback network, $\beta(s)$ and an adder that sums the input, X_i, and the feedback signal, X_f. The function of the feedback network is to sense the output, $X_o = H(s)X_e$, and produce the feedback signal,

$$X_f = \beta(s)X_o, \tag{2.1}$$

the adder output signal

$$X_e = X_i + \beta(s)X_o, \tag{2.2}$$

J. C. F. d. A. Casaleiro et al., *Quadrature RC–Oscillators*, Analog Circuits and Signal Processing, https://doi.org/10.1007/978-3-030-00740-9_2

Fig. 2.1 Oscillator feedback model

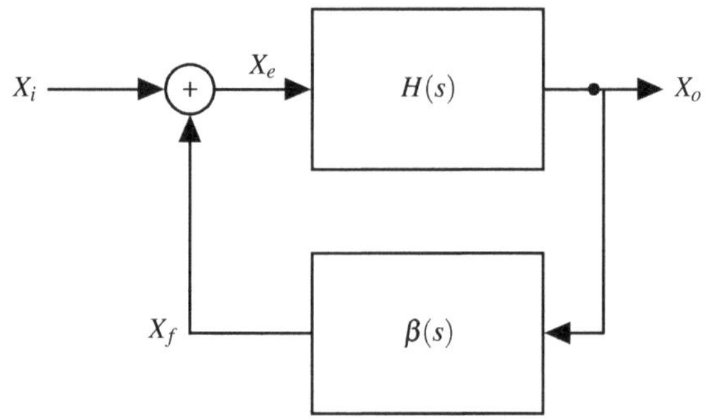

is applied to the forward network resulting in

$$X_o\,[1 - \beta(s)H(s)] = H(s)X_i. \tag{2.3}$$

An important aspect of Eq. (2.3) is that for zero input, $X_i = 0$, the output can be nonzero, if the left-hand side is zero, $1 - \beta(s)H(s) = 0$. For the case of oscillators, this particular case ($X_i = 0$) is known as the free-running mode, and the model of Eq. (2.3) is reduced to a closed loop including the forward and feedback networks. In the next chapters, we will discuss a more general case known as driven mode, where $X_i \neq 0$ and the input is used to couple or synchronize with other oscillators. From Eq. (2.3), we can derive the system transfer function:

$$A_f\,(s) = \frac{X_o(s)}{X_i(s)} = \frac{H(s)}{1 - H(s)\beta(s)} \tag{2.4}$$

For a steady-state oscillation to be maintained, the system poles must be purely imaginary, i.e. the equation $1 - \beta(s)H(s) = 0$ should have solutions with $s = \pm j\omega_0$, leading to the condition that the loop gain is $H(s)\beta(s) = 1$. This condition, known as the Barkhausen criterion, can be split into two conditions that must be met simultaneously. These two conditions concern the magnitude of the loop gain:

$$|H(s)\beta(s)| = 1, \tag{2.5}$$

and its phase:

$$\arg[H(s)\beta(s)] = 0 \tag{2.6}$$

To stabilize the oscillation frequency, the networks $H(s)$, $\beta(s)$ or both should be frequency-selective networks (resonators) that force the Barkhausen criterion to be met at a specific frequency, ω_0, as we will show in Sect. 2.4. An important aspect of the Barkhausen criterion is that it is a necessary, but not sufficient, condition for the oscillation to occur [1]. For instance, if we have a system with $\beta = 1$ and $|H(s)| > 1$, for any value of s, there is an exponential increase of the output, but no oscillation

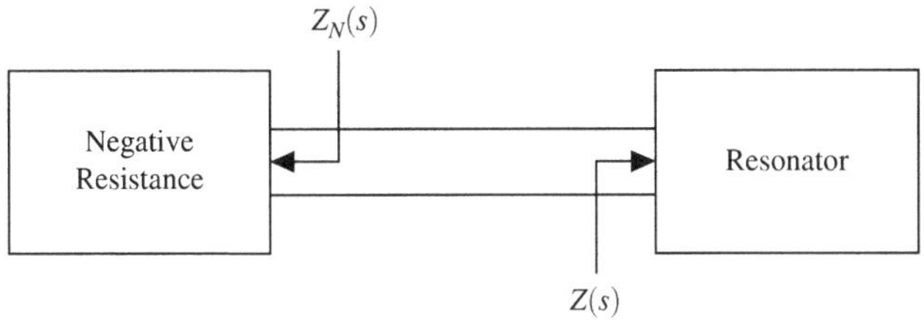

Fig. 2.2 Oscillator negative-resistance model

Fig. 2.3 Parallel LC-oscillator

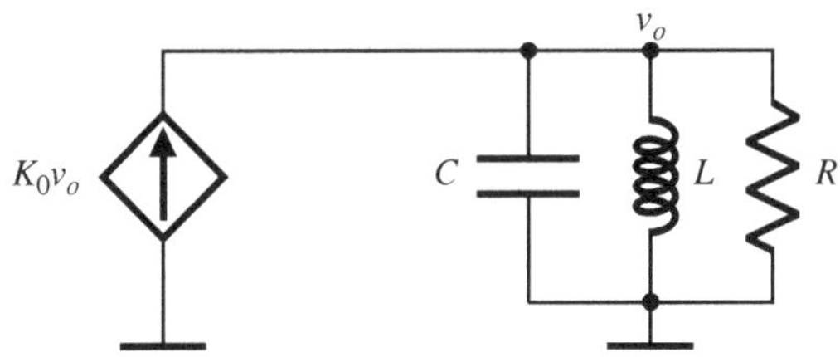

occurs, since there are no complex-conjugate poles [1]. Another example is at start-up, where the magnitude of the loop gain must be above unity $|H(s)\beta(s)| > 1$ [2]. For this reason, the oscillator loop gain is always designed slightly higher than one: the difference is known as excess loop gain. A loop gain higher than one will force the amplitude to grow, which is desirable at start-up, but it should be reduced to unity at steady state. This gain control mechanism, in the majority of oscillators, is due to nonlinearities, making the feedback model insufficient to analyse the amplitude stabilization, because assuming the system linearization.

An alternative model, described by Strauss in [3] and Kurokawa in [4], is the negative-resistance model, shown in Fig. 2.2, which represents the oscillator circuit as the connection of two one-port networks. The resonator is a frequency-selective network and defines the oscillation frequency. It can be made of passive or active elements. For example, in the parallel LC-oscillator model as in Fig. 2.3 one can easily set the negative resistance on the left, modeled by a voltage-dependent current source, and the passive resonator on the right.

Usually, in LC-oscillators the resonator is a passive network, and in RC-oscillators the resonator has active elements. In either case, the resonator is not lossless, with impedance $Z(\omega) = R + jX(\omega)$, which causes a fraction of the energy to be dissipated in the parasitic resistance. The equivalent impedance of the negative-resistance network is assumed to be $Z_N(A,\omega) = R(A,\omega) + jX(A,\omega)$. The impedance, Z_N, depends on the oscillation amplitude, A, due to the circuit nonlinearities.

To maintain the oscillation, the negative-resistance circuit must compensate the loss in R, leading to the steady-state oscillating condition $Z(\omega) = -Z_N(A,\omega)$. For the oscillation to start, the negative resistance should supply more energy than the loss in R. A negative-resistance behaviour can be obtained by using a nonlinear device, such as a tunnel diode, a Gunn diode or any device with a current-voltage

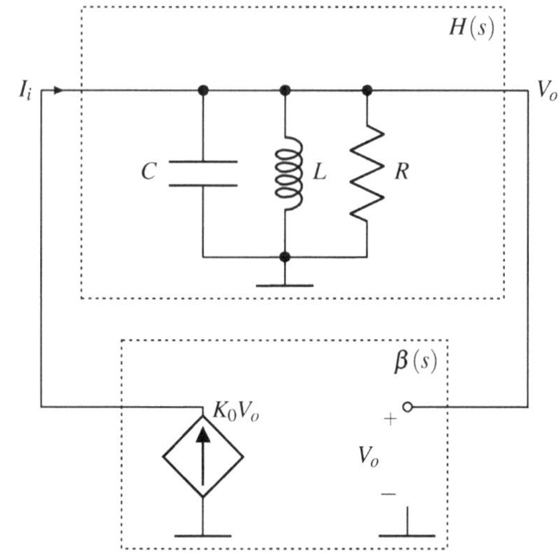

Fig. 2.4 Parallel LC-oscillator rearranged in a feedback model

characteristic having negative slope. It can also be based on an active circuit, as was just demonstrated in Fig. 2.3 and will be detailed along this book.

In the next sections, the LC-oscillator of Fig. 2.3 will be analysed using both models.

2.2.1 Feedback Model

Rearranging the circuit of Fig. 2.3 as shown in Fig. 2.4, one can write that the feedback transconductance, $\beta(s)$, is

$$\beta(s) = \frac{I_i}{V_o} = K_0, \tag{2.7}$$

and the transimpedance is

$$H(s) = \frac{V_o}{I_i} = \frac{s\frac{1}{RC}}{s^2 + s\frac{1}{RC} + \frac{1}{LC}} \tag{2.8}$$

Substituting Eq. (2.7) and Eq. (2.8) into Eq. (2.3), one obtains the characteristic equation:

$$s^2 + s\frac{1}{RC}(1 - K_0R) + \frac{1}{LC} = 0 \tag{2.9}$$

From Eq. (2.9), it is possible to obtain the condition for the loop gain:

$$K_0R \geq 1 \tag{2.10}$$

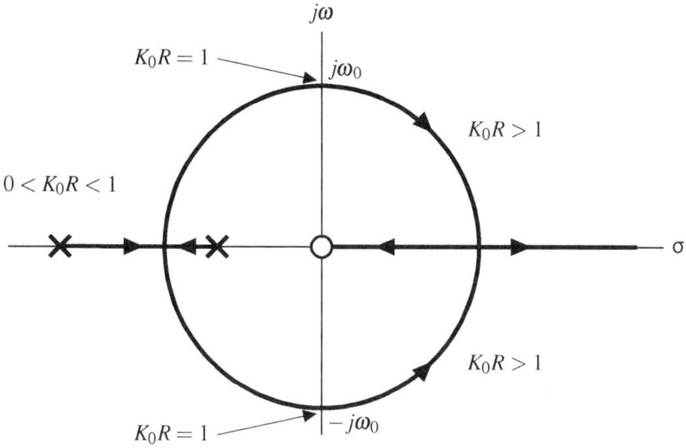

Fig. 2.5 Root locus

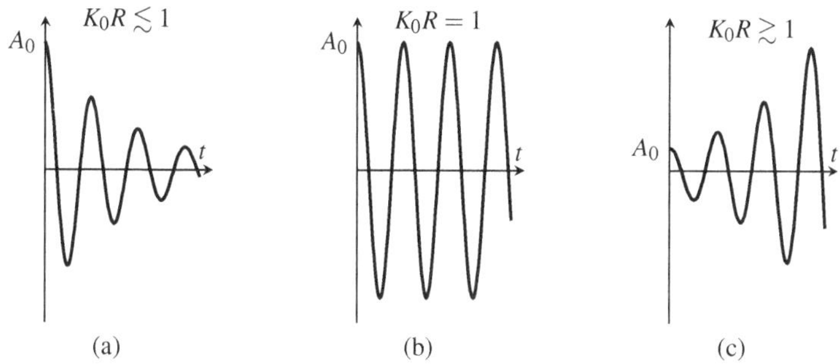

Fig. 2.6 Time solutions for amplitude: (**a**) decay, (**b**) steady and (**c**) growth

and the oscillation frequency:

$$\omega_0 = \frac{1}{\sqrt{LC}} \tag{2.11}$$

Substituting Eq. (2.7) and Eq. (2.8) into Eq. (2.4) and varying K_0, one can plot the root locus as shown in Fig. 2.5 and draw the same condition as in Eq. (2.10).

One can find such initial conditions that the time-domain solution of Eq. (2.9), for the loop gain near one, $K_0R_1 \lesssim 1$, is

$$v_o(t) \approx A_0 e^{-\frac{1-K_0R}{2RC}t} \cos(\omega_0 t), \tag{2.12}$$

where A_0 is the initial amplitude. Depending on K_0R, three possible particular solutions can be obtained, as shown in Fig. 2.6a, Fig. 2.6b, and Fig. 2.6c. For a loop gain slightly below unity, $K_0R_1 \lesssim 1$, the oscillation can start, but cannot

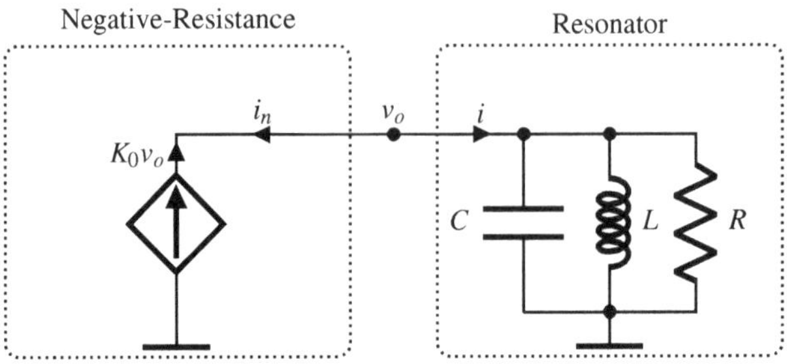

Fig. 2.7 Negative-resistance model of parallel LC-oscillator

be maintained because the amplitude will decay exponentially until the oscillation stops. For a loop gain equal to unity, $K_0R = 1$, and the loss in R is compensated, then the oscillation amplitude will be steady. For a loop gain with an excess, $K_0R_1 \gtrsim 1$, the oscillation amplitude will grow exponentially.

2.2.2 *Negative-Resistance Model*

The same circuit of Fig. 2.3 is analysed using the negative-resistance model.

Rearranging the circuit of Fig. 2.3 as shown in Fig. 2.7, one can obtain the resonator impedance which is given by:

$$Z(s) = \frac{V_o}{I} = \frac{s\frac{1}{C}}{s^2 + s\frac{1}{RC} + \frac{1}{LC}}, \tag{2.13}$$

and the negative impedance is

$$Z_N(s) = \frac{V_o}{I_n} = \frac{V_o}{-I} = -\frac{1}{K_0} \tag{2.14}$$

Applying Kurokawa's oscillation condition [4], $Z(s) = -Z_N(s)$, yields the same characteristic equation:

$$s^2 + s\frac{1}{RC}(1 - K_0R) + \frac{1}{LC} = 0 \tag{2.15}$$

Therefore, one can conclude that, for linear systems, both methods give the same result.

The dual circuit (Fig. 2.8) yields a similar result for the current, i. For clarity, we will refer to the circuit of Fig. 2.7 as the parallel LC-oscillator and to that of Fig. 2.8 as the series LC-oscillator. In the series LC-oscillator, the impedance of the resonator is

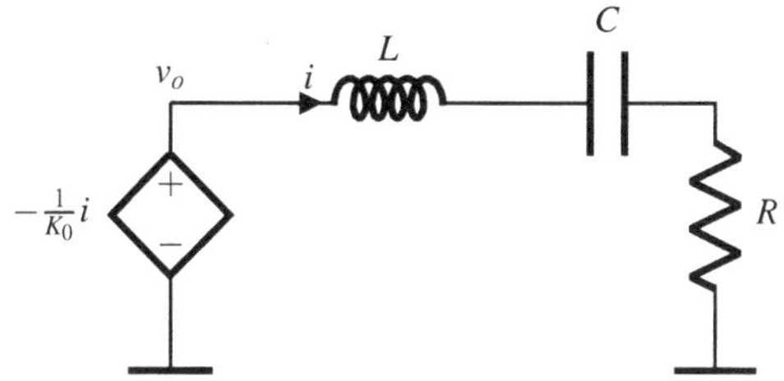

Fig. 2.8 Negative-resistance model of series LC-oscillator

$$Z(s) = \frac{V_o}{I} = sL + \frac{1}{sC} + R, \qquad (2.16)$$

and the impedance of the negative-resistance port is

$$Z_N(s) = \frac{V_o}{I_n} = \frac{V_o}{-I} = -\frac{1}{K_0} \qquad (2.17)$$

Applying again the Kurokawa's condition, one obtains the characteristic equation:

$$s^2 + s\frac{R}{L}\left(1 - \frac{1}{K_0 R}\right) + \frac{1}{LC} = 0, \qquad (2.18)$$

which yields the time-domain solution

$$i(t) = A_0 e^{-\frac{R}{2L}\left(1 - \frac{1}{K_0 R}\right)t} \cos \omega_0 t \qquad (2.19)$$

Equation (2.19) also shows that the oscillation starts when the system has an excess loop gain $K_0 R \geq 1$, similar to Eq. (2.16).

However, to reach a steady-state amplitude, an amplitude limiting technique is required in both cases to reduce the loop gain to one, $K_0 R = 1$, and to keep it at this value. In the next section, such techniques are discussed in detail.

2.2.3 *Negative-Resistance Circuits*

In modern oscillators, the negative resistance is often implemented by a cross-coupled differential pair, as shown in Fig. 2.9a. Here, we analyse the negative-resistance circuit of Fig. 2.9a for low frequency, resulting in the small-signal circuit shown in Fig. 2.9b. For the circuit (Fig. 2.9b), one can write

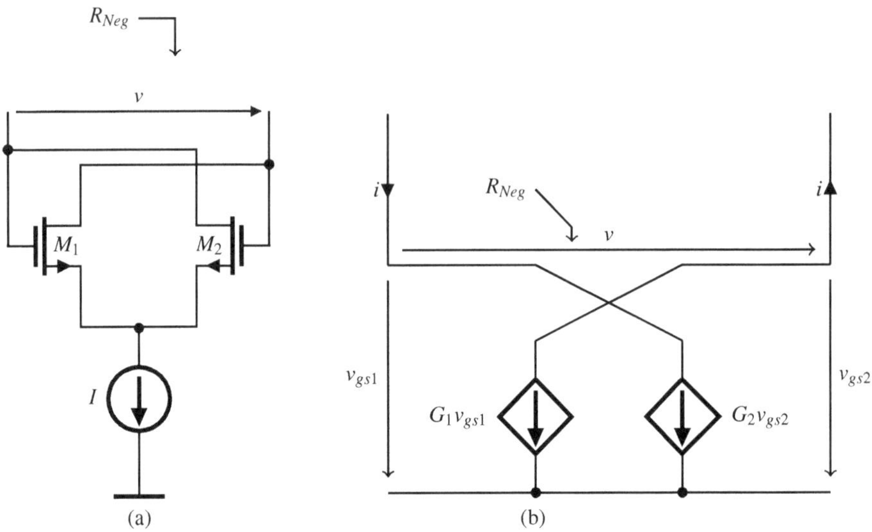

Fig. 2.9 Negative resistance circuit (**a**) and the small-signal equivalent (**b**)

$$\begin{cases} i = -g_{m1} v_{gs1} & \text{(2.20a)} \\ i = g_{m2} v_{gs2} & \text{(2.20b)} \\ v = v_{gs1} - v_{gs2} & \text{(2.20c)} \end{cases}$$

where g_{m1} and g_{m2} are the transistor's transconductances.

In case of perfect match, i.e. $g_{m1} = g_{m2} = g_m$, Eqs. (2.20) can be rewritten as:

$$\begin{cases} i = -g_m v_{gs1} & \text{(2.21a)} \\ i = g_m v_{gs2} & \text{(2.21b)} \\ v = v_{gs1} - v_{gs2} & \text{(2.21c)} \end{cases}$$

Dividing Eq. (2.21c) by the sum of Eq. (2.21a) with Eq. (2.21b), we obtain the negative resistance for the circuit of Fig. 2.9a:

$$R_{\text{Neg}} = \frac{v}{i} = -\frac{2}{g_m} \tag{2.22}$$

Another negative-resistance circuit often used is shown in Fig. 2.10a. From its small-signal equivalent circuit (Fig. 2.10b) and with matched transistors, one obtains

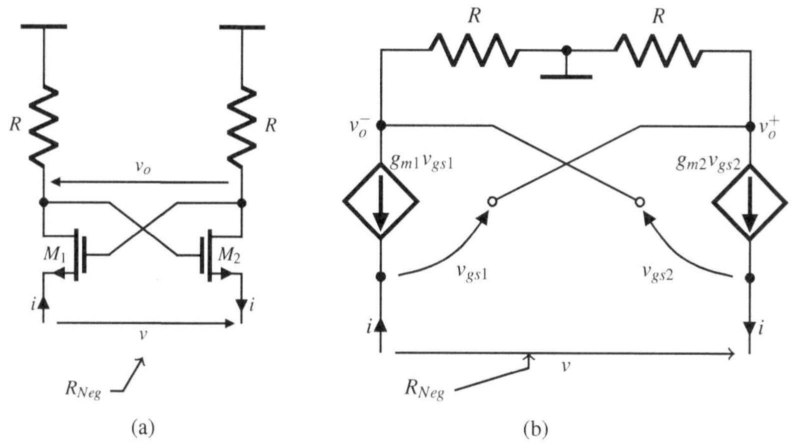

Fig. 2.10 Negative resistance circuit (**a**) and the small-signal equivalent (**b**)

$$\begin{cases} i = -g_m v_{gs1} & \text{(2.23a)} \\ i = g_m v_{gs2} & \text{(2.23b)} \\ v_o = v_o^+ - v_o^- = -2Ri & \text{(2.23c)} \\ v_o = \left(v_{gs1} - v_{gs2}\right) + v & \text{(2.23d)} \end{cases}$$

Substituting Eq. (2.23c) and the sum of Eq. (2.23a) with Eq. (2.23b) into Eq. (2.23d) and rearranging the equation results in the equivalent resistance of the circuit

$$R_{Neg} = \frac{v}{i} = \frac{2}{g_m} - 2R \tag{2.24}$$

From Eq. (2.24), we conclude that the equivalent resistance of the circuit (Fig. 2.10a) is a negative resistance in series with a positive resistance.

From the above analysis, we can conclude that these negative circuits for low amplitudes may compensate the losses.

2.3 Amplitude Limiting Techniques

Sinusoidal oscillators require an excess loop gain to ensure start-up, leading to an exponential growth of the oscillation amplitude. The amplitude must be regulated, or controlled, to avoid unwanted harmonics and distortion due to clipping. In this section, we review three methods to limit the oscillation amplitude.

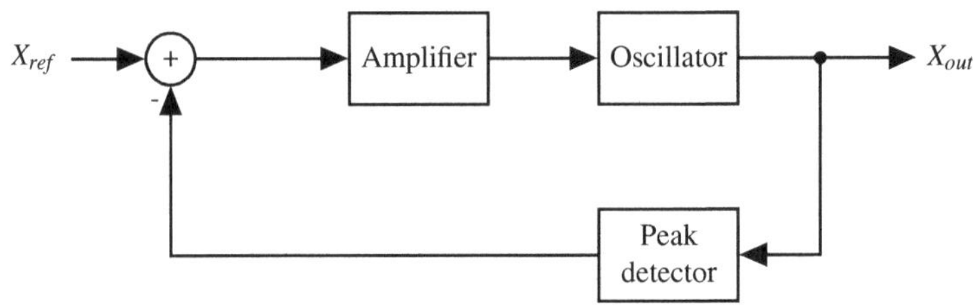

Fig. 2.11 Block diagram of an automatic amplitude control system

2.3.1 Automatic Amplitude Control

Using a feedback system to adjust the oscillator parameters and, therefore, maintain the amplitude steady is the most intuitive method.

In Fig. 2.11, a block diagram of an automatic amplitude control (AAC) based on measuring the amplitude is shown. This type of AAC circuit works as follows: the output amplitude is measured, using a peak detector, and the measured value is compared with a reference, X_{ref}. If the amplitude is larger than X_{ref}, the oscillator's gain is reduced and if the amplitude is lower than X_{ref}, the gain is increased, until the amplitude stabilizes at the reference level [5–7].

Using an AAC circuit, as the one of Fig. 2.11, has several advantages: it ensures a correct start-up; it allows to set the optimum gain to reduce the total harmonic distortion and minimize the phase noise [8]; it maintains a constant output power, regardless of the resonator quality factor, temperature and process variations [8]. The phase noise flattening, as reported in [9–11], is also an advantage. However, an automatic amplitude control (AAC) circuit increases the oscillator complexity, requiring more power and die area, and can become unstable [12, 13].

Another important aspect, often neglected, is to provide a possibility of independent amplitude and frequency controls [14]. Otherwise, a multi-input multi-output (MIMO) controller is required which increases the controller complexity. The *LC*-oscillator is an example in which the independence is guaranteed, since the frequency is determined only by the resonant tank, and the amplitude can be controlled by the negative resistance (usually controlled by the bias current).

However, usually this is not the case for *RC*-oscillators, which leads to a more complex controller, therefore limiting the use of this technique for this type of oscillators.

2.3.2 Piecewise-Linear Resistance

Another common method to limit the amplitude, usually for lower frequencies, is to use clamping elements (with nonlinear characteristics), e.g. rectifier diodes. This technique adds a discontinuity in the gain, or transconductance, of the oscillator's

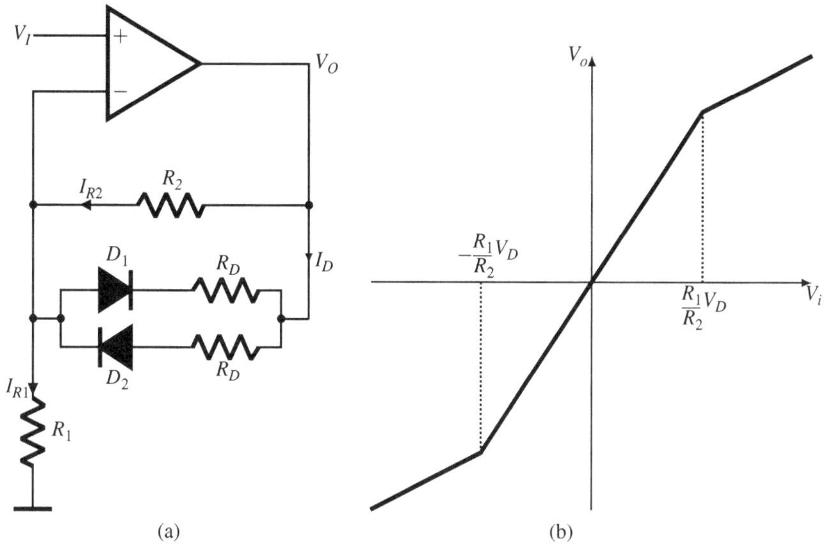

Fig. 2.12 Amplifier with nonlinear gain (**a**) circuit, and (**b**) transfer characteristic

active element. An example of this approach is the circuit shown in Fig. 2.12a, an amplifier with diodes in the feedback path. The amplifier has a high gain and the voltage between the amplifier input terminals is very small. To ease the analysis, let us assume that the diodes D_1 and D_2 have the same threshold voltage of Zener type and equal to V_D. If the voltage across R_2, $v_O - v_I$, is below this threshold voltage, both diodes are in cutoff and the amplifier's gain is given by:

$$A = \frac{dv_O}{dv_I} = \frac{v_o}{v_i} = \left(1 + \frac{R_2}{R_1}\right), \qquad \text{if} \quad |v_i| < \frac{R_1}{R_2} V_D \tag{2.25}$$

If the voltage across R_2 reaches the diodes' threshold voltage, $v_i \geq R_1 V_D / R_2$, then one of the diodes is conducting and the resistance R_D is connected in parallel with R_2. This reduces the gain to:

$$A = \frac{v_o}{v_i} \approx \left(1 + \frac{R_2//R_D}{R_1}\right). \tag{2.26}$$

Figure 2.12b shows the two slopes given by Eq. (2.25) and Eq. (2.26). Adjusting the slopes so that Eq. (2.25) gives the result high enough to start the oscillation and Eq. (2.26) low enough to stop the amplitude growth, the oscillation amplitude will be stabilized near the discontinuity point.

In comparison with the AAC method, this method leads to higher harmonic distortion [13] and depends on the ambient temperature due to the diode's threshold voltage dependency. Thus, to reduce the harmonic distortion, and to shape the output signal, more slopes can be introduced in the transfer function of the amplifier to smooth the transitions between slopes. An extensive study of this technique can be found in [3].

2.3.3 Inherent Circuit Nonlinearities

A third method is relying on the oscillator's intrinsic nonlinearities to limit the oscillation amplitude. In this case, the circuit parameters can be adjusted to obtain the desired output amplitude. For that, an accurate characterization of the oscillator is necessary, especially of its nonlinearities. The MOSFET transconductance for small signals is usually assumed to be constant with respect to the gate-source signal. However, for signals with moderate amplitude (hundreds of millivolts) this assumption is not valid. Here, we derive the MOSFET's signal-dependent transconductance equations for the strong and weak inversion operation regions.

2.3.3.1 Square-Law Nonlinearity

If the transistor is in strong inversion [15], the drain current is expressed as:

$$i_D = \frac{k}{2}\frac{W}{L}(v_{GS} - V_T)^2 \tag{2.27}$$

Here, k is a technology-dependent parameter, W and L are the transistor dimensions, v_{GS} is the gate-to-source voltage and V_T is the threshold voltage of the transistor. Assume that v_{GS} is

$$v_{GS} = V_{GS} + v_{gs}, \tag{2.28}$$

where V_{GS} is the bias and v_{gs} is the incremental component.

The MOSFET's drain current can be represented by a Taylor series expansion around the bias point, I_D, as:

$$i_D = I_D + \frac{di_D}{dv_{GS}}v_{gs} + \frac{1}{2}\frac{d^2 i_D}{dv_{GS}^2}v_{gs}^2 \tag{2.29}$$

Substituting Eq. (2.27) and Eq. (2.28) into Eq. (2.29), we obtain

$$i_D = I_D + k\frac{W}{L}(V_{GS} - V_T)v_{gs} + \frac{1}{2}k\frac{W}{L}v_{gs}^2 = I_D + g_{m0}v_{gs} + \frac{1}{2}k\frac{W}{L}v_{gs}^2 = I_D + G_m(v_{gs}) \tag{2.30}$$

where G_m is the signal-dependent transconductance given by:

$$G_m = g_{m0} + K_v v_{gs} \tag{2.31}$$

with $g_{m0} = k\frac{W}{L}\left(V_{GS} - V_T\right)$ and

$$K_v = \frac{1}{2}k\frac{W}{L} = \frac{g_{m0}^2}{4I_D} \tag{2.32}$$

Equation (2.31) shows that the transistor transconductance is linearly dependent on v_{gs}. Sometimes, it is convenient to have the transconductance as a function of the small-signal drain current. From Eq. (2.31), it follows that:

$$i_d = g_{m0}v_{gs} + K_v v_{gs}^2 = g_{m0}\left(1 + \frac{K_v}{g_{m0}v_{gs}}\right)v_{gs} \tag{2.33}$$

Assuming that $v_{gs} \approx (i_d/g_{m0})$ and substituting into Eq. (2.33),

$$i_d \approx g_{m0}\left(1 + \frac{K_v}{g_{m0}^2}i_d\right)v_{gs} \tag{2.34}$$

from which the current-dependent transconductance is obtained as:

$$G_m(i_d) \approx g_{m0}\left(1 + \frac{K}{g_{m0}^2}i_d\right) \tag{2.35}$$

2.3.3.2 Exponential Nonlinearity

If the transistor is in weak inversion [15], the drain current is expressed as:

$$i_D = I_{D0}\frac{W}{L}e^{\alpha(v_{GS}-V_T)}, \tag{2.36}$$

where I_{D0} is the technology parameter and α is an exponential factor given by:

$$\alpha = \frac{q}{nkT} \tag{2.37}$$

Here, q is the electron charge (1.6×10^{-19} C), k is Boltzmann's constant (1.38×10^{-23} J/K) and T is the absolute temperature in Kelvin. Using the Taylor approximation in Eq. (2.25), the large-signal drain current defined by Eq. (2.32) can be approximated as:

$$i_D = I_D + \alpha I_D v_{gs} + \frac{1}{2}\alpha^2 I_D v_{gs}^2 \tag{2.38}$$

From Eq. (2.38), we obtain the equation for the transconductance as a function of v_{gs}:

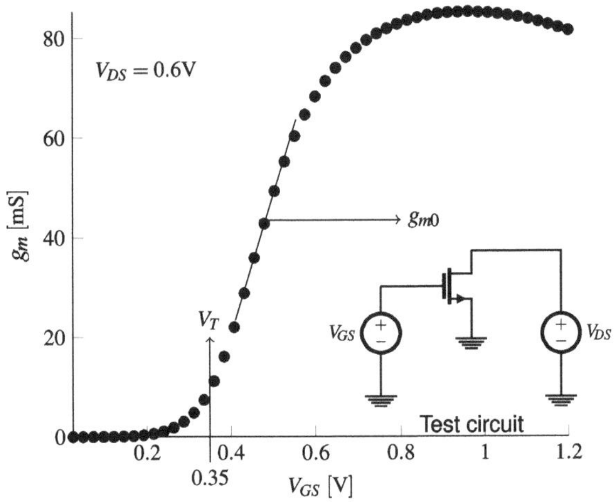

Fig. 2.13 Signal-dependent transconductance

$$G_m(v_{gs}) \approx g_{m0}\left(1 + \alpha v_{gs}\right) \tag{2.39}$$

where $g_{m0} = \alpha I_D$ is the transconductance.

To obtain the transconductance in terms of the incremental current, i_d, we assume that $v_{gs} \approx \frac{i_d}{g_{m0}}$. This results in

$$G_m(i_d) \approx g_{m0}\left(1 + \frac{\alpha}{g_{m0}} i_d\right) \tag{2.40}$$

Thus, it is possible to model the transistor's signal-dependent transconductance as a linear function of the incremental drain current, i_d, or incremental gate-source voltage, v_{gs}.

If the transistor is working in moderate inversion, a region between the strong and weak inversions, then for simplicity one can use the strong inversion model considering K as a fitting parameter.

To show the transconductance dependence of the gate voltage, the circuit shown in Fig. 2.13 was simulated using the parameters of UMC 130 nm CMOS technology. The transistor parameters are: 16 fingers with gate finger width of $W_f = 7.2\,\mu$m and gate length of $L = 120$ nm.

A DC sweep analysis of the V_{GS}, from 0 to 1.2 V, was done with $V_{DS} = 0.6$ V. The transistor transconductance, g_m, in relation to V_{GS}, is shown in Fig. 2.13. From Fig. 2.13, three distinct regions can be seen, the lower region, where V_{GS} is below the threshold voltage V_T, where the transistor is working in the sub-threshold region. In the middle region, the transistor is working in strong inversion, and g_m increases almost linearly with V_{GS} where Eq. (2.31) is valid. In the region for $V_{GS} > 0.6$ V, the transconductance, g_m, can be approximated by a constant value around 81.7 mS.

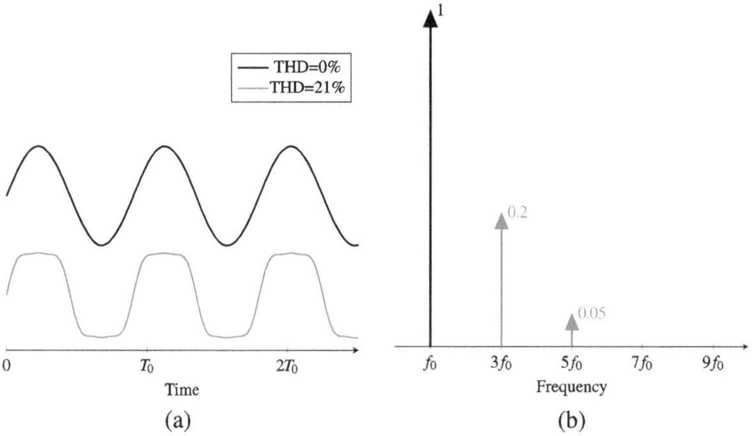

Fig. 2.14 Comparison of a sinusoid and a distorted signal (**a**) in time, and (**b**) in frequency domain

The most linear region of g_m, assuming a maximum relative error of 5%, can be found around $V_{GS} = 0.468$ V. It can be linearized with relative error of 4.52%. The Eq. (2.31) parameters are $K = 0.2795$ and $g_{m_0} \approx 43.5$ mS.

2.3.4 Harmonic Distortion

An ideal sinusoidal oscillator generates an output signal with a single frequency, f_0, which is referred to as the fundamental, as shown in Fig. 2.14a, top waveform. In practical sinusoidal oscillators, the amplitude limiting and the circuit nonlinearities distort the output signal. These distortions are described by harmonics, frequency components which have frequencies nf_0, i.e. the multiples of the fundamental frequency. The total harmonic distortion (THD) is the figure of merit used to quantify the level of harmonics in a signal. As an example, it is shown in the time (Fig. 2.14a) and frequency (Fig. 2.14b) domains a comparison of a sinusoidal signal (THD = 0%) and a distorted signal with two harmonics with 20% and 5% amplitude, respectively, of the fundamental f_0 which result in a THD = 21%.

There are two definitions for the THD. The definition mostly used for RF oscillators is given by:

$$\mathrm{THD} = \frac{\sqrt{\sum_{n=2}^{\infty} V_n^2}}{V_1} \tag{2.41}$$

where V_n is the amplitude of the harmonics, and V_1 is the amplitude of the fundamental. It compares the high-frequency harmonic content of the oscillator's output signal with its fundamental.

A second definition given by:

$$\mathrm{THD} = \sqrt{\frac{\sum_{n=2}^{\infty} V_n^2}{\sum_{n=1}^{\infty} V_n^2}} \tag{2.42}$$

is mostly used in audio systems, and compares the high-frequency harmonic contents of the output signal with its rms value. Both definitions give nearly the same result for low values of THD, i.e. THD below 1%. An excellent review of THD can be found in [16].

2.4 Frequency Selectivity

The oscillation frequency of a sinusoidal oscillator is set by the resonator. From the feedback model perspective, the resonator makes the required phase shift so that the loop gain phase be 2π at the oscillation frequency. In the negative-resistance model, the resonator is a bandpass filter that attenuates unwanted frequencies passing only its resonance frequency, ω_0. Both cases describe the so-called free-running oscillator, with an oscillation frequency equal to the resonant frequency, or close to it if we include the circuit's parasitic elements. However, that is not the case for coupled oscillators, as it will be shown in the following chapters.

2.4.1 Quality Factor

The opposition of an oscillator to any deviation from its natural oscillation frequency is quantitatively defined by the Q-factor. Hence, an oscillator with a high Q will have a more stable oscillation frequency since it strongly opposes to any deviation from its oscillation frequency. An oscillator with a low Q will have a less stable oscillation frequency. Usually, an oscillator deviates from its natural frequency due to circuit nonlinearities and noise. Next, we will consider different definitions of Q and discuss their equivalence. In the literature [1, 2], a common definition of Q is

$$Q = 2\pi \frac{\text{Maximum Energy stored}}{\text{Energy dissipated per oscillation cycle}}, \tag{2.43}$$

which physically means the number of oscillation cycles that a resonator does with the maximum energy stored and not replenished in the subsequent oscillations. From this definition, we can derive the Q of a resonator network. For the parallel RLC resonator, the voltage is the same for all elements. Hence, the maximum energy stored per cycle is related to the maximum voltage, V_{peak}, at the capacitor, so that

$$E_C = \frac{1}{2} C V_{\text{peak}}^2 \tag{2.44}$$

where V_{peak} is the oscillation amplitude (peak voltage). The energy dissipated per cycle in the resistor is

$$E_R = \int_0^T P(t)\mathrm{d}t = \int_0^T \frac{v^2(t)}{R}\mathrm{d}t, \tag{2.45}$$

where $v(t) = V_{\text{peak}} \cos(\omega_0 t)$, T is the oscillation period and R is the resistance value. Using the trigonometric identity $\cos \omega_0 t^2 = 1/2\,(1 + \cos(2\omega_0 t))$, we get

$$E_R = \frac{1}{2R} V_{\text{peak}}^2 T. \tag{2.46}$$

Substituting Eq. (2.44) and Eq. (2.46) into Eq. (2.43) results in the well-known quality factor of the parallel RLC circuit:

$$Q = 2\pi \frac{\frac{1}{2} C V_{\text{peak}}^2}{\frac{1}{2R} V_{\text{peak}}^2 T} = \omega_0 RC = R\sqrt{\frac{C}{L}}. \tag{2.47}$$

For the series RLC resonator, it is the current that is common to all elements. Hence, the maximum energy stored per cycle is related to the maximum current in the inductor:

$$E_L = \frac{1}{2} L I_{\text{peak}}^2, \tag{2.48}$$

where I_{peak} is the amplitude of the current. The energy dissipated per cycle in the resistor is

$$E_R = \int_0^T R i^2(t)\mathrm{d}t = \frac{1}{2} R I_{\text{peak}}^2 T, \tag{2.49}$$

where $i(t) = I_{\text{peak}} \cos(\omega_0 t)$.

Substituting Eq. (2.48) and Eq. (2.49) into Eq. (2.43) results in

$$Q = 2\pi \frac{\frac{1}{2} L I_{\text{peak}}^2}{\frac{1}{2} R I_{\text{peak}}^2 T} = \omega_0 \frac{L}{R} = \frac{1}{R}\sqrt{\frac{L}{C}}. \tag{2.50}$$

To use equations Eq. (2.47) and Eq. (2.50), the exact circuit parameters: R, L and C must be known. In practice, this is not always possible due to parasitic elements. However, if the exact Q cannot be determined this way, then it can be determined by the method proposed by Leeson in [17]. This method defines Q of the resonator as:

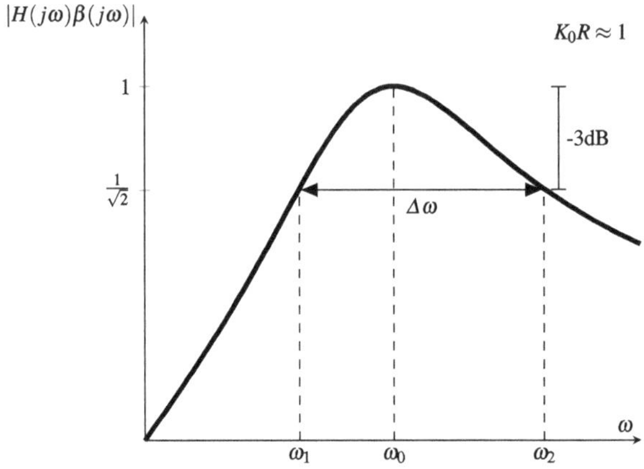

Fig. 2.15 Loop gain frequency response

$$Q = \frac{\omega_0}{\Delta\omega} \tag{2.51}$$

where ω_0 is the resonant frequency and $\Delta\omega$ is the -3-dB bandwidth. This definition allows to measure the Q from the resonator's frequency response. No formal proof was presented in [17]. The proof that both definitions are equivalent was derived in [18]. Take the loop gain of the LC-oscillator of Sect. 2.2, that we rewrite here for clarity, and assume $K_0R = 1$:

$$H(s)\beta(s) = \frac{s\frac{1}{RC}}{s^2 + s\frac{1}{RC} + \frac{1}{LC}}. \tag{2.52}$$

Then, the -3-dB attenuation of the loop gain, as shown in Fig. 2.15, is obtained from:

$$|H(j\omega)\beta(j\omega)|^2 = \frac{\omega^2\left(\frac{1}{RC}\right)^2}{\left(\omega_0^2 - \omega^2\right)^2 + \omega^2\left(\frac{1}{RC}\right)^2} = \frac{1}{2}. \tag{2.53}$$

Solving for ω, we obtain the polynomial:

$$\omega^2 \pm \omega\frac{1}{RC} - \omega_0^2 = 0 \tag{2.54}$$

with the positive roots:

$$\omega_1 = -\frac{1}{2RC} + \sqrt{\left(\frac{1}{RC}\right)^2 + 4\omega_0^2},\ \omega_2 = \frac{1}{2RC} + \sqrt{\left(\frac{1}{RC}\right)^2 + 4\omega_0^2}. \tag{2.55}$$

Subtracting the roots, we obtain the -3-dB bandwidth:

$$\Delta\omega = \omega_2 - \omega_1 = \frac{1}{RC}. \tag{2.56}$$

Substituting Eq. (2.56) into Eq. (2.51) for $\omega = \omega_0$ results in:

$$Q = \frac{\omega_0}{\frac{1}{RC}} = R\sqrt{\frac{C}{L}} \tag{2.57}$$

which is the same as Eq. (2.47). Hence, we can conclude that both definitions yield the same result for second-order resonators [2]. A similar conclusion can be drawn for the series RLC resonator. However, for oscillators with distributed elements, which cannot be reduced to a second-order RLC circuit, the definition given by Eq. (2.57) is not accurate, as explained in [18, 19].

A third definition can be found in [20] and in [21]. It is based on the feedback model and defines Q in terms of the slope of the phase characteristics at the resonance frequency:

$$Q = -\frac{1}{2}\omega_0\frac{\partial\theta}{\partial\omega} \tag{2.58}$$

Afterwards, it was shown that this definition can be applied only to oscillators with resonators, since it only considers the phase. The definition fails for resonator-less oscillators like the two-integrator and the phase-shift oscillator [18].

A fourth definition proposed in [18], called the open-loop Q, is based on the open-loop gain derivatives of the magnitude and phase,

$$Q = \frac{\omega_0}{2}\sqrt{\left(\frac{\mathrm{d}A}{\mathrm{d}\omega}\right)^2 + \left(\frac{\mathrm{d}\theta}{\mathrm{d}\omega}\right)^2} \tag{2.59}$$

where A is the magnitude and θ is the phase of the loop gain. This definition is especially useful for analysis using the feedback model. A similar definition was proposed in [22], using the phase slope or group delay to determine the quality factor. Where the S-parameters are used to describe the open loop gain. More recently, the definition proposed in [18] was extended in [23] and generalized to one- and two-port passive networks and in [24] to active networks as well. The Q factor is defined in [24] as the derivative of the port impedance logarithm,

$$Q = \frac{\omega_0}{2}\left|\frac{\mathrm{d}}{\mathrm{d}\omega}\ln\left(Z(j\omega)\right)\right| = \frac{\omega_0}{2}\frac{1}{|Z|}\left|\frac{\mathrm{d}Z}{\mathrm{d}\omega}\right|, \tag{2.60}$$

where Z is the resonator impedance. Using the resonator's impedance presented in Sect. 2.2, we can verify the equivalence between the definition of [24] and other four definitions. Starting with the series RLC resonator, we know that the impedance is

$$Z(s) = sL + \frac{1}{sC} + R. \tag{2.61}$$

For $s = j\omega$, the magnitude of the impedance is

$$|Z(j\omega)| = \sqrt{\left(\omega L - \frac{1}{\omega C}\right)^2 + R^2} \tag{2.62}$$

and the derivative is

$$\left|\frac{\mathrm{d}Z(j\omega)}{\mathrm{d}\omega}\right| = \left(L + \frac{1}{\omega^2 C}\right) = \left(\frac{\omega^2 LC + 1}{\omega^2 C}\right). \tag{2.63}$$

Substituting Eq. (2.62) and Eq. (2.63) into Eq. (2.60) with $\omega = \omega_0$ results in

$$Q = \frac{\omega_0}{2}\frac{1}{R}\frac{2}{\omega_0^2 C} = \frac{1}{R}\sqrt{\frac{L}{C}}. \tag{2.64}$$

For the parallel RLC, the impedance is

$$Z(s) = \frac{s\frac{1}{C}}{s^2 + s\frac{1}{RC} + \omega_0^2}, \tag{2.65}$$

For $s = j\omega$, the magnitude of the impedance is

$$|Z(j\omega)| = \frac{\omega\frac{1}{C}}{\sqrt{\left(\omega_0^2 - \omega^2\right)^2 + \omega^2\left(\frac{1}{RC}\right)^2}}, \tag{2.66}$$

and the derivative is

$$\left|\frac{\mathrm{d}Z(j\omega)}{\mathrm{d}\omega}\right| = 2R^2 C. \tag{2.67}$$

Substituting Eq. (2.66) and Eq. (2.67) into Eq. (2.60) with $\omega = \omega_0$ results in

$$Q = \frac{\omega_0}{2}\frac{1}{R}2R^2 C = R\sqrt{\frac{C}{L}}. \tag{2.68}$$

Hence, this definition yields the same result for second-order resonators, and therefore, we can conclude that all definitions are equivalent.

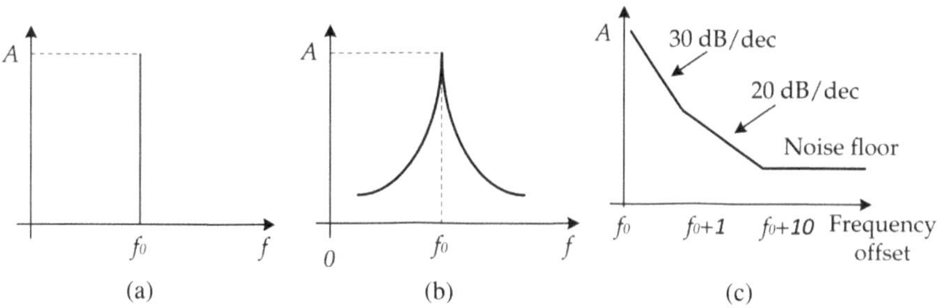

Fig. 2.16 Spectrum of the output signal of (**a**) a noiseless oscillator, (**b**) an oscillator with phase noise, (**c**) in single sideband asymptotic graph

2.4.2 *Phase Noise*

The phase noise and Q are the most common parameters used to express the merit of oscillators. Here, we will discuss briefly the phase noise; a comprehensive discussion can be found in [1, 2, 18, 25].

An ideal sinusoidal oscillator generates a spectrum with a single component, as shown in Fig. 2.16a. The noise in oscillators has a modulating effect, generating a signal modulated in amplitude and in frequency. This noise can be either external or generated internally in the circuit. Usually, external noise comes from the power supply.

The external noise can be reduced to negligible levels, with proper design of the power supply, mainly, using low-noise voltage regulators and high-order filtering. Thus, we focus on the internal noise. The internal noise is generated by the circuit elements. Within the circuit, several noise sources contribute to the overall noise. The active elements contribute with flicker, shot and thermal noise. Resistors contribute with thermal noise, and inductors and capacitors do not generate noise. The noise is not uniformly distributed near the resonant frequency; instead, there is a specific spectrum shape, as shown in Fig. 2.16b. The effect of noise in the time domain is shown in Fig. 2.17.

2.4.2.1 Leeson's Model

The phase noise spectrum shape [17] is expressed by:

$$L\left(\omega_m\right) = 10\log\left[\left(\frac{2FkT}{P}\right)\left[1+\left(\frac{\omega_0}{2Q\omega_m}\right)^2\right]\left(1+\frac{\omega_{\text{flicker}}}{\omega_m}\right)\right] \tag{2.69}$$

where k is the Boltzmann constant, T is the absolute temperature, P is the average power dissipated in the tank, ω_0 is the oscillation frequency, ω_m is the offset frequency, ω_{flicker} is the corner frequency of the flicker noise zone, Q is the quality

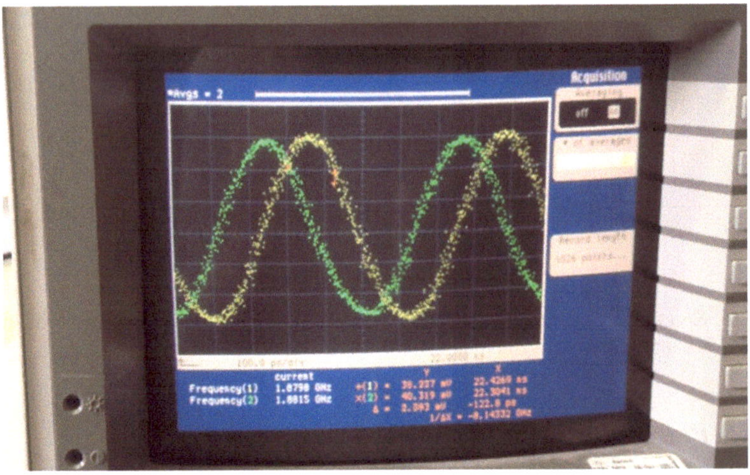

Fig. 2.17 Output of quadrature oscillator with noise

factor of the tank and F is the excess noise factor, which is an empirical parameter. From Eq. (2.69), three separated zones can be found. For offset frequencies far from the oscillation frequency, i.e. $\omega_m \gg \omega_0$ and $\omega_m \gg \omega_{\text{flicker}}$, there is a constant noise floor given by:

$$L_{\text{white noise}} \approx 10 \log \frac{2FkT}{P} \tag{2.70}$$

Due to the frequency modulation of the noise, a −20-dB/decade slope exists for moderate offset frequencies, given by:

$$L_{\text{FM noise}} \approx 20 \log \left(\frac{\omega_0}{2Q\omega_m} \right) \tag{2.71}$$

The slope increases to −30 dB/decade for offset frequencies $\omega_m \leq \omega_{\text{flicker}}$, due to the flicker noise.

The Leeson's model considers the oscillator as a linear and time-invariant system. A more accurate model considers the oscillator as a linear time variant (LTV) system [26]. The LTV model in the time domain was first proposed by Hajimiri and Lee in [27], and later in the frequency domain by Samori et al. [28], Huang [29] and Rael and Abidi [30]. The LTV model gives interesting insights concerning the circuit parameters. Using this model, several research works derived the minimum theoretical achievable noise, for several oscillator topologies [26–28, 30–32].

References

1. B. Razavi, *RF Microelectronics* (Prentice-Hall, Upper Saddle River, 1998)
2. L. Oliveira, J. Fernandes, I. Filanovsky, C. Verhoeven, M. Silva, *Analysis and Design of Quadrature Oscillators* (Springer, Heidelberg, 2008)
3. L. Strauss, *Wave Generation and Shaping* (McGraw-Hill, New York, 1970)
4. K. Kurokawa, *An Introduction to the Theory of Microwave Circuits* (Academic, New York, 1969)
5. D. Meyer-Ebrecht, Fast amplitude control of a harmonic oscillator. Proc. IEEE **60**(6), 736–736 (1972)
6. E. Vannerson, K.C. Smith, Fast amplitude stabilization of an RC oscillator. IEEE J. Solid State Circuits **9**(4), 176–179 (1974)
7. I.M. Filanovsky, Oscillators with amplitude control by restoration of capacitor initial conditions. IEE Proc. G Electron. Circuits Syst. **134**(1), 31–37 (1987)
8. M.A. Margarit, J.L. Tham, R.G. Meyer, M.J. Deen, A low-noise, low-power VCO with automatic amplitude control for wireless applications. IEEE J. Solid State Circuits **34**(6), 761–771 (1999)
9. A. Zanchi, C. Samori, A. Lacaita, S. Levantino, Impact of AAC design on phase noise performance of VCOs. IEEE Trans. Circuits Syst. II **48**(6), 537–547 (2001)
10. J. Rogers, D. Rahn, C. Plett, A study of digital and analog automatic-amplitude control circuitry for voltage-controlled oscillators. IEEE J. Solid State Circuits **38**(2), 352–356 (2003)
11. O. Casha, I. Grech, J. Micallef, E. Gatt, Design of a 1.2 V low phase noise 1.6 GHz CMOS buffered quadrature output VCO with automatic amplitude control, in *13th IEEE International Conference on Electronics, Circuits and Systems* (2006), pp. 192–195
12. B. Linares-Barranco, T. Serrano-Gotarredona, A loss control feedback loop for VCO stable amplitude tuning of RF integrated filters, in *IEEE International Symposium on Circuits and Systems*, vol. 1 (2002), pp. 521–524
13. T. O'Dell, Instability of an oscillator amplitude control system. IEE Proc. Control Theory Appl. **151**(2), 194–197 (2004)
14. I.M. Filanovsky, A sinusoidal VCO with control of frequency and amplitude, in *Proceedings of the 32nd Midwest Symposium on Circuits and Systems*, vol. 1 (1989), pp. 446–449
15. D.A. Johns, K. Martin, *Analog Integrated Circuit Design*, 2nd edn. (Wiley, New York, 2011)
16. D. Shmilovitz, On the definition of total harmonic distortion and its effect on measurement interpretation. IEEE Trans. Power Delivery **20**(1), 526–528 (2005)
17. D. Leeson, A simple model of feedback oscillator noise spectrum. Proc. IEEE **54**(2), 329–330 (1966)
18. B. Razavi, A study of phase noise in CMOS oscillators. IEEE J. Solid State Circuits **31**(3), 331–343 (1996)
19. J.-C. Nallatamby, M. Prigent, M. Camiade, J. Obregon, Extension of the Leeson formula to phase noise calculation in transistor oscillators with complex tanks. IEEE Trans. Microw. Theory Tech. **51**(3), 690–696 (2003)
20. K. Clarke, D. Hess, *Communication Circuits: Analysis and Design* (Addison-Wesley, New York, 1971)
21. R.W. Rhea, *Oscillator Design and Computer Simulation* (Prentice-Hall, Upper Saddle River, 1990), p. 47
22. M. Randall, T. Hock, General oscillator characterization using linear open-loop S-parameters. IEEE Trans. Microw. Theory Tech. **49**(6), 1094–1100 (2001)
23. T. Ohira, Rigorous Q-factor formulation for one- and two-port passive linear networks from an oscillator noise spectrum viewpoint. IEEE Trans. Circuits Syst. II **52**(12), 846–850 (2005)
24. T. Ohira, K. Araki, Active Q-factor and equilibrium stability formulation for sinusoidal oscillators. IEEE Trans. Circuits Syst. II **54**(9), 810–814 (2007)

25. L.B. Oliveira, A. Allam, I.M. Filanovsky, J.R. Fernandes, C.J.M. Verhoeven, M.M. Silva, Experimental comparison of phase-noise in cross-coupled RC- and LC-oscillators. Int. J. Circuit Theory Appl. **38**, 681–688 (2010)
26. D. Murphy, J.J. Rael, A.A. Abidi, Phase noise in LC oscillators: a phasor-based analysis of a general result and of loaded Q. IEEE Trans. Circuits Syst. I Regul. Pap. **57**(6), 1187–1203 (2010)
27. A. Hajimiri, T. Lee, A general theory of phase noise in electrical oscillators. IEEE J. Solid State Circuits **33**(2), 179–194 (1998)
28. C. Samori, A.L. Lacaita, F. Villa, F. Zappa, Spectrum folding and phase noise in LC tuned oscillators. IEEE Trans. Circuits Syst. II Analog Digit. Signal Process. **45**(7), 781–790 (1998)
29. Q. Huang, Phase noise to carrier ratio in LC oscillators. IEEE Trans. Circuits Syst. I Fundam. Theory Appl. **47**(7), 965–980 (2000)
30. J.J. Rael, A.A. Abidi, Physical processes of phase noise in differential LC oscillators, in *Proceedings of the IEEE 2000 Custom Integrated Circuits Conference* (2000), pp. 569–572
31. P. Andreani, X. Wang, L. Vandi, A. Fard, A study of phase noise in Colpitts and LC-tank CMOS oscillators. IEEE J. Solid State Circuits **40**(5), 1107–1118 (2005)
32. P. Andreani, A. Fard, More on the $1/f^2$ phase noise performance of CMOS differential-pair LC-tank oscillators. IEEE J. Solid State Circuits **41**(12), 2703–2712 (2006)

Chapter 3
Van der Pol Oscillator

3.1 Introduction

The van der Pol oscillator model is an approximation valid for a fairly wide class of oscillators. The literature dedicated to calculation of the waveforms in this oscillator is very extensive; references [1–7] are only a small portion of it. This oscillator was the first analytically investigated oscillator, and, afterwards, was used as a model for studying nonlinear oscillations in electronic systems.

In practice, this is the only oscillator for which the oscillator harmonics distortion was calculated analytically. Using the harmonics amplitudes, one can find a very important performance parameter, the total harmonic distortion (THD). Even though the VDPO is only an approximation to real oscillators, the calculations using this model give a reasonable estimation of THD, as will be shown in this chapter.

3.2 Van der Pol Model

A classic example of the VDPO is the parallel LC-oscillator with a nonlinearity to limit the amplitude growth, as shown in Fig. 3.1. Using the one-port model, the circuit is divided into a negative-resistance block at the left-hand side and a resonator at the right-hand side. The negative-resistance block has a linear term, $K_0 v_o$, and a nonlinear term, $K_2 v_o^3$. The first generates the negative resistance to compensate the resonator's loss, and the latter regulates the amplitude growth.

Since the circuit of Fig. 3.1 has a nonlinearity, it is simpler to derive the system dynamics in the time domain, rather than working in the frequency domain. Thus, applying the Kirchhoff's current law at the output node, we obtain

$$C\frac{\mathrm{d}v_o}{\mathrm{d}t} + \frac{1}{L}\int v_o \mathrm{d}t + \frac{v_o}{R} = K_0 v_o - K_2 v_o^3, \tag{3.1}$$

J. C. F. d. A. Casaleiro et al., *Quadrature RC–Oscillators*, Analog Circuits and Signal Processing, https://doi.org/10.1007/978-3-030-00740-9_3

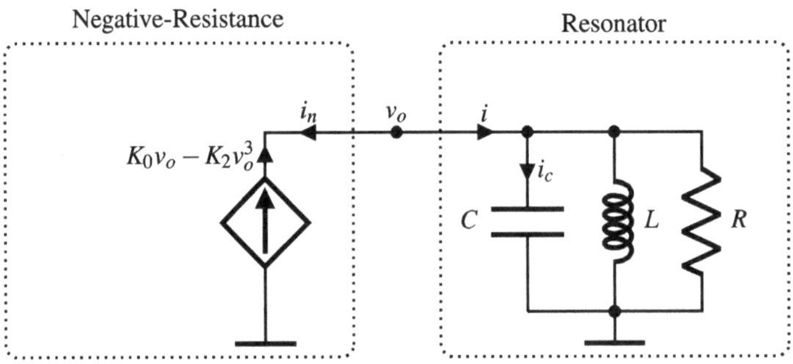

Fig. 3.1 Parallel van der Pol oscillator

which by differentiating and dividing both sides by the capacitance C can be expressed as a nonlinear second-order differential equation:

$$\frac{d^2v_o}{dt^2} + \frac{1}{RC}\left[(1 - K_0R) + 3K_2Rv_o^2\right]\frac{dv_o}{dt} + \frac{1}{LC}v_o = 0. \tag{3.2}$$

Equation (3.2) can be reduced to the general form of the VDP equation:

$$\frac{d^2v_o}{dt^2} - 2\left(\delta_0 - \delta_2 v_o^2\right)\frac{dv_o}{dt} + \omega_0^2 v_o = 0, \tag{3.3}$$

where $\delta_0 = (K_0R - 1)/(2RC)$ and $\delta_2 = 3K_2R/(2RC)$ represent the negative resistance and the amplitude limiter, respectively, and ω_0 is the oscillator free-running frequency.

To assess the stability of the oscillator, we take the Eq. (3.3) and convert it to the equivalent system of two first-order differential equations:

$$\begin{cases} \dfrac{dv_o}{dt} = \dfrac{1}{C}i_c & (3.4a) \\ \dfrac{di_c}{dt} = 2C\left(\delta_0 - \delta_2 v_o^2\right) i_c - C\omega_0^2 v_o & (3.4b) \end{cases}$$

in terms of the output voltage, v_o, and the capacitor current, i_c.

From Eqs. (3.4), we can determine the equilibrium points, also known as critical points, by equating the derivatives to zero. In this case, there is a single equilibrium point at $i_c = 0$ and $v_o = 0$. An equilibrium point is where the system can stay in equilibrium (assuming a noiseless system). From the mathematical point of view, this means that a zero amplitude is a solution of Eq. (3.3). Physically, it means that if the oscillator starts with the capacitor discharged and the inductor voltage and current zero it will remain in that state permanently.

We start by studying the local stability near the equilibrium point based on the linearized system. We find the linear version of the system, for any point in the phase

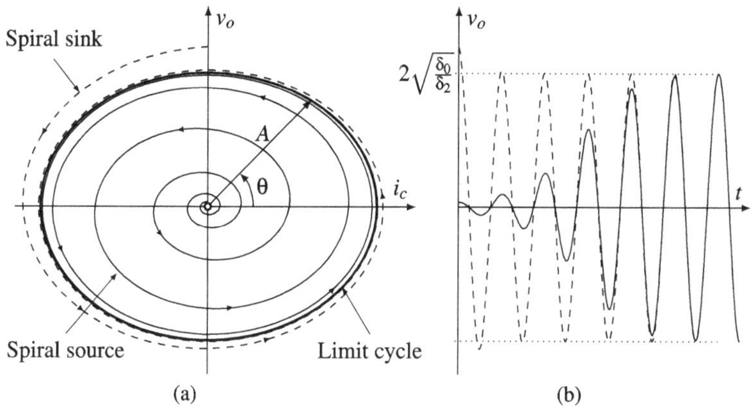

Fig. 3.2 Negative-resistance circuit (**a**) and the small-signal equivalent (**b**)

space, by calculating its Jacobian matrix:

$$J = \begin{bmatrix} \frac{\partial f}{\partial v_o} & \frac{\partial f}{\partial i_c} \\ \frac{\partial g}{\partial v_o} & \frac{\partial g}{\partial i_c} \end{bmatrix} \Rightarrow \begin{bmatrix} 0 & \frac{1}{C} \\ -4\delta_2 C i_c v_o - C\omega_0^2 & 2C\left(\delta_0 - \delta_2 v_o^2\right) \end{bmatrix}, \tag{3.5}$$

where f is a function equal to the derivative of the output voltage, $f = \mathrm{d}v_o/\mathrm{d}t$, and g is a function equal to the derivative of the capacitance current, $g = \mathrm{d}i_c/\mathrm{d}t$.

From the eigenvalues of the Jacobian matrix, we can determine the stability [1]. The eigenvalues are the roots of the characteristic polynomial:

$$|J - \lambda I| = 0 \Rightarrow \begin{vmatrix} 0 - \lambda & \frac{1}{C} \\ -4\delta_2 C i_c v_o - C\omega_0^2 & 2C\left(\delta_0 - \delta_2 v_o^2\right) - \lambda \end{vmatrix} = 0, \tag{3.6}$$

where λ is an eigenvalue and I is the identity matrix. From Eq. (3.6), we obtain the characteristic polynomial:

$$\lambda^2 - T\lambda + D = 0, \tag{3.7}$$

where T is the trace of J and D is the determinant of J. For the VDPO, the trace is

$$T = 2C\left(\delta_0 - \delta_2 v_o^2\right), \tag{3.8}$$

and the determinant is

$$D = \left(4\delta_2 i_c v_o + \omega_0^2\right). \tag{3.9}$$

The stability conditions are: $T < 0$ and $D > 0$. For the equilibrium point at the origin, we have

$$D(v_o = 0, i_c = 0) = \omega_0^2 > 0, \tag{3.10}$$

and

$$T(v_o = 0, i_c = 0) = 2C\delta_0 > 0, \tag{3.11}$$

This means that the equilibrium point is unstable. Since $\delta_0 \sim (K_0 R - 1)$, we can say that the oscillation starts if $K_0 R > 1$ (the equilibrium point must be unstable). Furthermore, we can say that, near the equilibrium point, we have a spiral source, as shown in the phase diagram of Fig. 3.2a, because $T^2 < 4D$, resulting in two complex-conjugate eigenvalues. We can check this condition by solving Eq. (3.7) for λ, resulting in

$$\lambda = \frac{T}{2} \pm \frac{1}{2}\sqrt{T^2 - 4D}, \tag{3.12}$$

Since the system is an oscillator, we expect the existence of a stable limit cycle.[1] The existence of a stable limit cycle indicates that at some point, far from the equilibrium point, a spiral sink must exist outside the limit cycle. Complex-conjugate eigenvalues with $T < 0$ is a necessary condition for the existence of a spiral sink. From Eq. (3.8), it is clear that if, and only if, the output voltage is $v_o > \sqrt{\delta_0/\delta_2}$, $T < 0$ and, therefore, a spiral sink and a limit cycle exist.

Based on the qualitative analysis of the VDPO, without explicitly determining the solution, we can conclude that the system has an unstable equilibrium point at the origin and a stable limit cycle so the oscillation amplitude is limited, as shown in Fig. 3.2a. Since a stable limit cycle exists, the solution is a sinusoidal signal as shown in Fig. 3.2b.

We can make a simpler qualitative analysis, assuming that the output signal, v_o, is sinusoidal

$$v_o(t) = A(t)\cos(\omega_0 t + \phi). \tag{3.13}$$

Next, we use the harmonic balance method, presented in [1], to obtain the oscillation amplitude and frequency. We included in Appendix A a detailed description of an alternative method presented by Theodorchik in [8].

Using the harmonic balance method presented in [1], we substitute Eq. (3.13) into Eq. (3.3) the amplitude and phase derivatives are preliminary obtained. Assuming a slow variation of the amplitude, we obtain

[1]A limit cycle is a closed trajectory in the phase diagram that represents a periodic solution.

$$\begin{cases} \dfrac{dA}{dt} = \delta_0 A - \dfrac{1}{4}\delta_2 A^3 & (3.14a) \\ \dfrac{d\phi}{dt} = 0, & (3.14b) \end{cases}$$

Equation (3.14b) means that any constant phase satisfies the second equation. The phase value will depend on the initial conditions of the circuit and will be maintained indefinitely in steady state.

For the amplitude, there are three equilibrium points: $A = 0$, $A = 2\sqrt{\delta_0/\delta_2}$ and $A = -2\sqrt{\delta_0/\delta_2}$ as can be determined from Eq. (3.14). However, we will consider only positive values for the amplitude, since the negative values can be represented by the choice of phase $\phi = \pi$. The plot of Eq. (3.14a) is shown in Fig. 3.3. A qualitative analysis of Eq. (3.14a), illustrated by Fig. 3.3 shows that the equilibrium point $A = 0$ is unstable, because $dA/dt > 0$ for $A > 0$, meaning that the oscillator can start with a zero amplitude, but for any deviation from the equilibrium point the amplitude will increase. The second equilibrium point $A = 2\sqrt{\delta_0/\delta_2}$ is an attractor, because for an amplitude below this equilibrium point $dA/dt > 0$, and for amplitudes above that equilibrium point $dA/dt > 0$. We also know that this attractor is stable because it is related to the stable limit cycle determined before (see Fig. 3.2a).

Equation (3.14a) can also be used to obtain the analytical solution for the amplitude, since it is a separable first-order equation [1]. The general solution is

$$\int \frac{1}{\delta_0 A - \frac{1}{4}\delta_2 A^3} dA = \int dt - T_0,$$

where T_0 is an arbitrary constant. Integrating gives

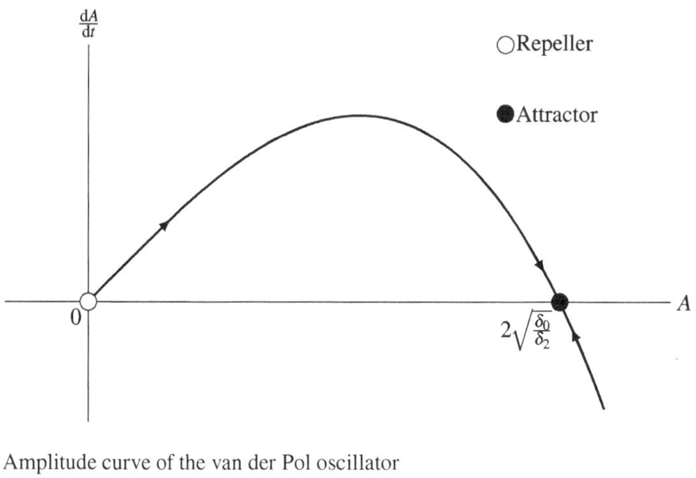

Fig. 3.3 Amplitude curve of the van der Pol oscillator

$$\frac{1}{2\delta_0}\ln\left(\frac{4A^2}{4\delta_0 - \delta_2 A^2}\right) = t - T_0,$$

from which the transient amplitude response results as:

$$A(t) = 2\sqrt{\frac{\delta_0}{\delta_2 + A_0 e^{-2\delta_0 t}}}, \tag{3.15}$$

where $A_0 = 4\delta_0^{-1} e^{2\delta_0 T_0}$ is a parameter that defines the initial amplitude of the circuit, for $t = 0$.

For the steady state, when $t \to \infty$, the oscillation amplitude given by Eq. (3.15) is reduced to:

$$A = 2\sqrt{\frac{\delta_0}{\delta_2}}. \tag{3.16}$$

The complete solution for the parallel topology of the VDPO with $K_0 R > 1$ is obtained by substituting Eq. (3.15) into Eq. (3.13), resulting in:

$$v_o(t) = 2\sqrt{\frac{\delta_0}{\delta_2 + A_0 e^{-2\delta_0 t}}}\cos(\omega_0 t + \phi).$$

A similar solution for the series topology can be also obtained, but the result is expressed in terms of the current and not the voltage.

3.3 Practical Realization of a van der Pol Oscillator

We discuss and evaluate the harmonics content in the considered oscillator. This is a very old problem [1–6], which did not lose importance till now [8–21]. Many mathematical methods of the nonlinear oscillations analysis were created (an excellent review of the early work can be found in [13]), and this work is still going on. Some recent work [14–17] is well adapted to electronic oscillators in general; other methods [18, 19] are directed to the analysis of specific oscillators. Methods of low-distortion oscillator design continue to proliferate [20, 21].

THD is a major performance parameter for quasi-sinusoidal oscillators. In the usual practice, an oscillator is designed first, then it is simulated or tested, and the design is verified for THD. The design is accepted if the THD satisfies the performance requirements. Otherwise, the oscillator is redesigned (and here the above mentioned mathematical methods are finally used) or some measures of THD reduction are applied. In this chapter, we show that it is possible to set a certain level of THD at the start of design. This is valid for the oscillators which can be described by the VDP model. Fortunately, as was mentioned above, this is a relatively wide class of oscillators.

3.3.1 Oscillator Characteristic Equation

The RLC-oscillator of Fig. 3.4a is analysed assuming mismatches only in the transistors' transconductances. We describe here how the oscillator can be represented by the corresponding VDP model. We give the amplitude calculations, compare the results of calculations and simulations and discuss the difference in the results.

The incremental model of the circuit is shown in Fig. 3.4b. Adding all parasitic capacitances into C_T results in:

$$C_T = C + 2C_{gd} + \frac{C_{gs}}{2}$$

In steady-state oscillation, the transistors' transconductances G_1 and G_2 are functions of the gate-source voltages of M_1 and M_2 (see Sect. 2.3), and this relation is important for the amplitude stabilization, assuming, at this stage, that there are no mismatches and that the signal is antisymmetric, i.e. $v_{gs1} = -v_{gs2} = v_o/2$, we have antisymmetric signal-dependent transconductances:

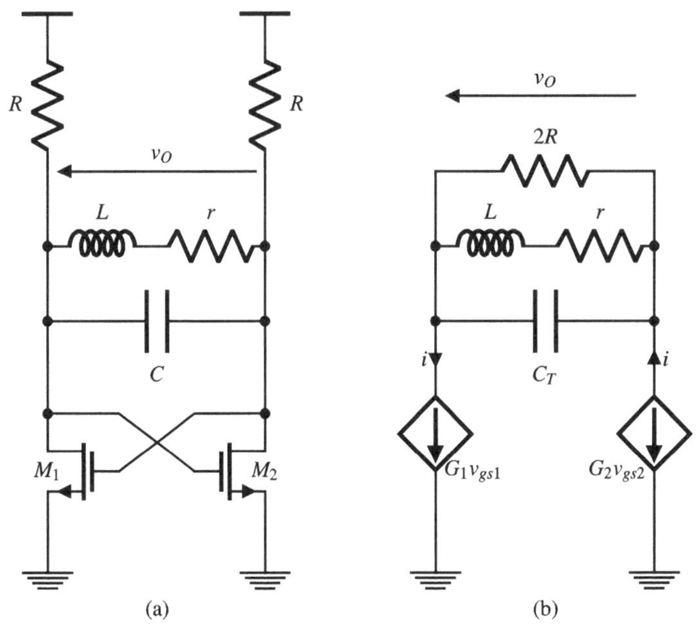

Fig. 3.4 MOS RLC-oscillator circuit (**a**) and incremental model (**b**)

$$\begin{cases} G_1 = g_{m0} + K\dfrac{v_o}{2} & (3.17a) \\ G_2 = g_{m0} - K\dfrac{v_o}{2} & (3.17b) \end{cases}$$

Applying the KCL to the circuit of Fig. 3.4b, we obtain

$$\begin{cases} i = G_1 v_{gs1} & (3.18a) \\ i = -G_2 v_{gs2} & (3.18b) \\ v_o = v_{gs1} - v_{gs2} \quad . & (3.18c) \end{cases}$$

Substituting Eq. (3.18a) and Eq. (3.18b) into Eq. (3.18c), using the signal dependent transconductances, given by Eq. (3.17a) and Eq. (3.17b), and solving with respect to the current, i, we obtain the current as a function of the output voltage:

$$i = \frac{G_1 G_2}{G_2 + G_1} v_o = \frac{g_{m0}}{2} v_o - \frac{K^2}{8 g_{m0}} v_o^3. \tag{3.19}$$

The second term on the right-hand side of Eq. (3.19) indicates a significant distortion for high-amplitude input signals. However, for small amplitudes this term can be neglected, resulting in:

$$i \approx \frac{g_{m0}}{2} v_o. \tag{3.20}$$

Thus, for small amplitude, the $i(v)$ characteristic is almost linear, as shown in Fig. 3.5. The figure shows a comparison between the theory and simulation using transistors' models of a standard CMOS technology. Note that if the amplitude of current i is equal to half of the bias current, then transistor M_1 is in strong inversion, but M_2 is in cutoff. Conversely, if the amplitude current i is equal to minus half of the bias current, then transistor M_2 is in strong inversion, but M_1 is in cutoff. Thus, Eq. (3.19) and Eq. (3.20) are valid for $|v_o| < I_{\text{bias}}/g_{m0}$. This validity region is indicated at the top of Fig. 3.5.

3.3.2 *Transition to van der Pol Equation*

Equation (3.19) shows that the transistors can be modeled by a negative resistance in parallel with a nonlinear current source. Using Eq. (3.19), we can redraw the incremental circuit, as shown in Fig. 3.6. Applying the Kirchhoff voltage and current laws to the circuit of Fig. 3.6, we obtain

$$C_T \frac{d^2 v}{dt^2} + \frac{v}{2R} + \frac{v}{R_N} + K v^3 + \frac{1}{L}\int v dt - \frac{r}{L}\int i_l dt = 0. \tag{3.21}$$

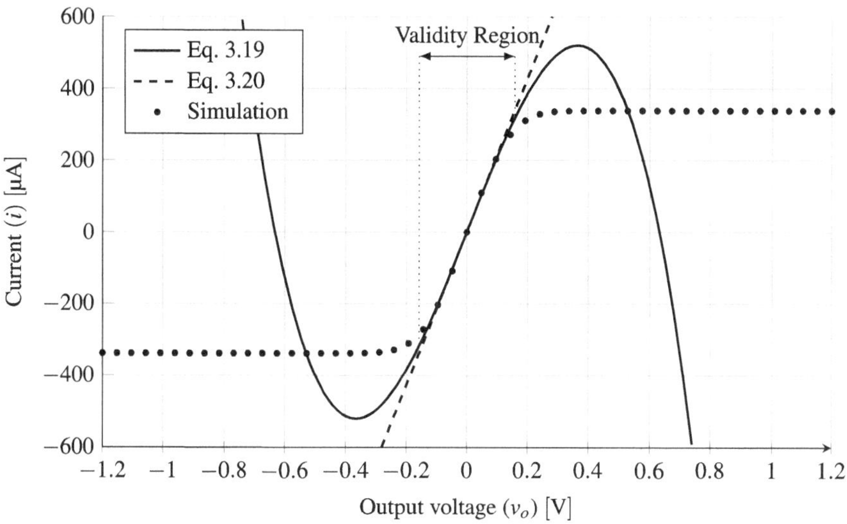

Fig. 3.5 Current of the cross-coupled transistors as a function of the differential voltage. Transistor dimensions are $W = 14.4\,\mu\text{m}$, $L = 120\,\text{nm}$, $I_{\text{bias}} = 676\,\mu\text{A}$ and $g_{m0} = 4.28\,\text{mS}$

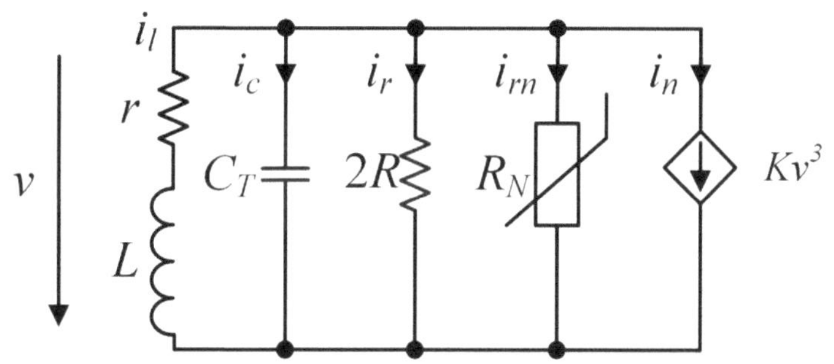

Fig. 3.6 Incremental model of MOS RLC-oscillator

where $R_N = -2/g_{m0}$ is the negative resistance, $K = 2k/8g_{m0}$ is a parameter of the nonlinear term and i_l is the incremental current of the inductance L. Differentiating Eq. (3.21) and dividing it by C_T, one obtains

$$\frac{d^2v}{dt^2} + \frac{1}{C_T}\left[\frac{1}{2R} + \frac{1}{R_N} + 3Kv^2\right]\frac{dv}{dt} + \frac{1}{LC_T}v = \frac{r}{LC_T}i_l. \tag{3.22}$$

To eliminate i_l from Eq. (3.22), we substitute the right-hand side by the sum of the other currents, namely,

$$i_l = -\left(i_r + i_c + i_{rn} + i_n\right) = -\frac{v}{2R} - \frac{v}{2R_N} - Kv^3 - C_T\frac{dv}{dt}$$

This will give us

$$\frac{d^2v}{dt^2}+\left[\frac{r}{L}+\frac{1}{2C_T}\left(\frac{1}{R}-\frac{g_{m0}}{R}+6Kv^2\right)\right]\frac{dv}{dt}+\omega_0 v=0. \tag{3.23}$$

where

$$\omega_0\approx\sqrt{\frac{2+r\left(R^{-1}-g_{m0}\right)}{2LC_T}} \tag{3.24}$$

As follows from Eq. (3.23), that to start the oscillations one has to set

$$g_{m0}>\frac{1}{R}+\frac{2C_T r}{L} \tag{3.25}$$

Equation (3.24) shows that the losses in r should be also compensated (which is usually ignored). This is very important in case of inductors with high losses (integrated inductors are the best example).

Finally, using the notation:

$$\delta_0=\frac{1}{2}\left(\frac{Rg_{m0}-1}{2RC_T}-\frac{r}{L}\right);\qquad \delta_2=\frac{g_{mn}}{4C_T} \tag{3.26}$$

one can rewrite Eq. (3.23) as:

$$\frac{d^2v}{dt^2}-2\left(\delta_0-\delta_2v^2\right)\frac{dv}{dt}+\omega_0^2v=0. \tag{3.27}$$

This is the familiar form of the van der Pol's equation. The solution of Eq. (3.27), as derived in Sect. 3.2, is given by:

$$v=A_{10}\sin\omega_0 t \tag{3.28}$$

where:

$$A_{10}=2\sqrt{\frac{\delta_0}{\delta_2}}=2\sqrt{1-\frac{1}{g_{m0}}\left(\frac{1}{R}+\frac{2rC_T}{L}\right)} \tag{3.29}$$

Equation (3.28) is valid for small amplitudes. In practical oscillators of this type, the amplitude A_{10}, as it follows from Eq. (3.29), quickly becomes close to its asymptotic value of $A_{10}\approx 2$. But, as one will see in the next section, this increase of amplitude involves the development of harmonics. The oscillation frequency also becomes different from the fundamental frequency ω_0.

3.3.3 THD Calculation

Using the normalized variable $y = v/\sqrt{\delta_0/\delta_2}$, and the natural time $\tau = \omega_0 t$, one can rewrite Eq. (3.27) as:

$$\frac{d^2y}{d\tau^2} - \lambda\left(1 - y^2\right)\frac{dy}{d\tau} + y = 0. \tag{3.30}$$

Here:

$$\lambda = \left(\frac{g_{mn}R - 1}{2RC_T} - \frac{r}{L}\right)\sqrt{\frac{2LC_T}{2 - (r/R)(g_{mn}R - 1)}} \tag{3.31}$$

This normalized form of van der Pol equation usually appears in the mathematical literature. It was well investigated, numerically [3, 4] and then analytically [5–7]. We use below the results of [5], because the calculations are done using simple algebra, and the covered range of λ is sufficient for integrated oscillators. More general results can be found in [6, 7]. It was shown in [5] that for $0 < \lambda < 3$ the solution of Eq. (3.30) can be expressed as:

$$\begin{aligned} y = {} & A_1 \cos \Omega\tau + A_3 \cos 3\Omega\tau + B_3 \sin 3\Omega\tau + A_5 \cos 5\Omega\tau + B_5 \sin 5\Omega\tau \\ & + A_7 \cos \Omega\tau + B_7 \cos 7\Omega\tau \ldots \end{aligned} \tag{3.32}$$

where each amplitude has two components, namely,

$$\begin{cases} A_1 = A_{10} + A_{12} & (3.33a) \\ A_3 = A_{30} + A_{32}; \quad B_3 = B_{30} + B_{32} & (3.33b) \\ A_5 = A_{51} + A_{52}; \quad B_5 = B_{51} + A_{52} & (3.33c) \\ A_7 = A_{71} + A_{72}; \quad B_7 = B_{71} + A_{72} & (3.33d) \end{cases}$$

The oscillation frequency, Ω, also consists of two components, namely,

$$\Omega = \Omega_0 + \Omega_2 \tag{3.34}$$

However, for calculation of amplitude components it is necessary to find first

$$\Omega_0 = \sqrt{\frac{4 - \lambda^2 + \sqrt{\lambda^4 + 16}}{8}} \tag{3.35}$$

Then, one has to find the first components of the amplitudes:

$$B_{30} \approx -\frac{2\lambda\Omega_0}{\lambda^2 + 8\Omega_0^2}; \quad A_{30} \approx \frac{\lambda B_{30}}{\Omega_0} \tag{3.36}$$

and only then to calculate

$$A_{10} \approx 2 - \frac{A_{30}}{2} - \frac{B_{30}^2}{2} \tag{3.37}$$

We put these components in the order convenient for calculations especially by writing a MATLAB script. After that, one can calculate the components:

$$\begin{aligned} A_{51} &\approx \frac{\lambda\Omega_0 B_{30}}{5}; \qquad B_{51} \approx -\frac{2\lambda^2 B_{30}}{75}; \\ A_{71} &\approx \frac{11\lambda^2 B_{30}^2}{68\Omega_0^2}; \qquad B_{71} \approx -\frac{\lambda B_{30}^2\left(49\Omega_0^2 - 14\lambda^2\right)}{686\Omega_0^3} \end{aligned} \tag{3.38}$$

Now, one can calculate the corrections to these components, first we obtain

$$\begin{aligned} A_{12} &\approx \frac{\lambda^2 B_{30}^2}{25}; \quad A_{32} \approx \frac{\lambda\left[2B_{30}\left(1 - 9\Omega_0^2\right) + 6\lambda\Omega_0\right]}{6\Omega_0\left(\lambda^2 + 8\Omega_0^2\right)} A_{51}; \\ B_{32} &\approx \frac{\lambda\left(1 - 9\Omega_0^2 - 6\lambda\Omega_0 B_{30}\right)}{3\left(\lambda^2 + 8\Omega_0^2\right)} A_{51} \end{aligned} \tag{3.39}$$

Finally, to finish these calculations, one has to find

$$\begin{aligned} A_{52} &\approx -\frac{2\lambda^2 A_{51}}{\lambda^2 + 24\Omega_0^2}; \qquad B_{52} \approx \frac{10\lambda\Omega_0 A_{51}}{\lambda^2 + 24\Omega_0^2}; \\ A_{72} &\approx -\frac{\lambda A_{51}}{7\Omega_0\left(\lambda^2 + 48\right)}; \qquad B_{72} \approx -\frac{7\lambda\Omega_0 A_{51}}{7\Omega_0\left(\lambda^2 + 48\right)} \end{aligned} \tag{3.40}$$

For calculation of THD, we have to find the amplitudes $C_1 = A_1$ and $C_i = \sqrt{A_i^2 + B_i^2} (i = 3, 5, 7)$, from Eqs. (3.33). All these amplitudes are functions of λ only. The results of these calculations are given in Fig. 3.7.

The amplitude normalization does not influence the harmonic amplitude ratio and these harmonics $C_i (i = 1, 3, 5, 7)$ can be directly used for an approximate calculation of THD (in percentage) as:

$$\text{THD} \approx \frac{\sqrt{C_3^2 + C_5^2 + C_7^2}}{C_1} \times 100 \tag{3.41}$$

To finish this section, we mention that the correction of frequency:

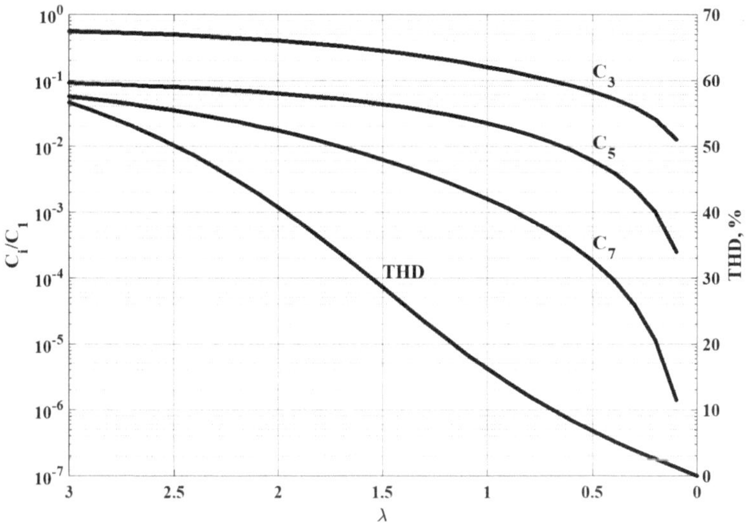

Fig. 3.7 Van der Pol oscillator harmonic amplitudes versus λ

$$\Omega_2 \approx -\frac{4\lambda^4\Omega_0^5}{10\left(1+\Omega_0^2\right)\left(8\Omega_0^2+\lambda^2\right)^2}\left(1+\frac{2\lambda^2}{75\Omega_0^2}\right) \tag{3.42}$$

is very small in the considered range of λ, and the result Eq. (3.35) is sufficient to evaluate the reduction of the oscillation frequency which becomes

$$\omega \approx \Omega_0\omega_0 \tag{3.43}$$

that is, it will be less than the natural frequency ω_0.

3.3.4 Oscillator Design Example

The theoretical results of the two previous sections were verified by simulating the oscillator (Fig. 3.4a) designed in a 130-nm digital CMOS process. The initial calculations assumed that $\mu_n C_{OX} = 367\,\mu\text{A/V}^2$, $V_{TN} = 0.301\,\text{V}$, $C_{OX} = 13\,\text{fF}/\mu\text{m}$, $\gamma_n = 0.28\,\text{V}^{-1}$ and $2\phi_f = 0.85\,\text{V}$. Here, we used the extrapolated data for submicron technologies given in [22]. The oscillator was designed for the supply voltage of 1.2 V. The coil has the inductance of $L = 44\,\text{nH}$, with $r = 20\,\Omega$. The transistors have the aspect ratios $W/L = 47.6/0.13\,\mu\text{m}/\mu\text{m}$. The bias resistor is $R = 100\,\Omega$. Using this data and the process parameters, one finds from the large-signal model of the transistor that $I = 6.01\,\text{mA}$. One also finds that

$g_m = 40.2\,\text{mS}$, $C = 53.6\,\text{fF}$ and $C_{gd} = C_{gs}/4$. Simulation of the static operation gives $I = 5.94\,\text{mA}$, $g_{m0} = 32.2\,\text{mS}$, $C_{gs} = 47.3\,\text{fF}$ and $G_{gd} = 12.9\,\text{fF}$.

Figure 3.8 shows the waveforms for large $C = 6\,\text{pF}$ and small $C = 150\,\text{fF}$ values of the tank capacitor (Fig. 3.4a). One can see that for large values of C the waveforms are close to sinusoidal. Yet, the reduction of tank capacitor results in strong distortions which are not taken into consideration by a simple linear theory developed in the first parts of this chapter. One can also notice that the amplitude of this distorted wave is higher than the amplitude of the sinusoidal wave.

Figure 3.9 gives the comparison of calculated and simulated amplitudes and THD as a function of tank capacitance C. The simulated amplitudes are always smaller than the calculated ones. This is explained by the fact that the approximation of nonlinearity in a practical oscillator is different from that required by the VDPO theory. In addition, the difference between calculated and simulated amplitudes increases for smaller values of the tank capacitor. The THD quickly increases when, by reducing C, one tries to design an oscillator operating at higher frequency. The predicted THD is somewhat pessimistic (the figure is approximately 1.5 times higher than that obtained in simulations). Thus, the increase of harmonic content is closely connected with decrease of frequency with respect to the result provided by linear theory. Figure 3.10 gives the comparison of the frequency given by a linear model, f_0 obtained from Eq. (3.24), nonlinear model f obtained from Eq. (3.43) and the frequency f_{sim} obtained in simulations.

One can see that the reduction of frequency predicted by nonlinear mode is rather pessimistic. But, the predicted and simulated results in the high-frequency region

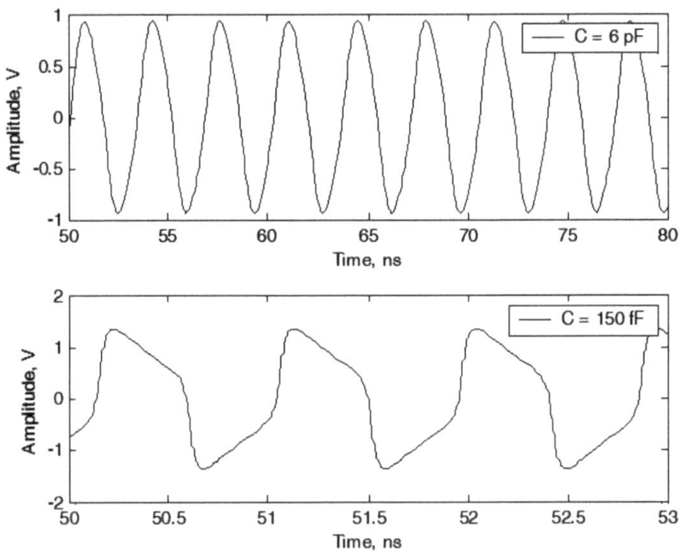

Fig. 3.8 Simulation results for large (6 pF) and small (150 fF) values of the tank capacitance

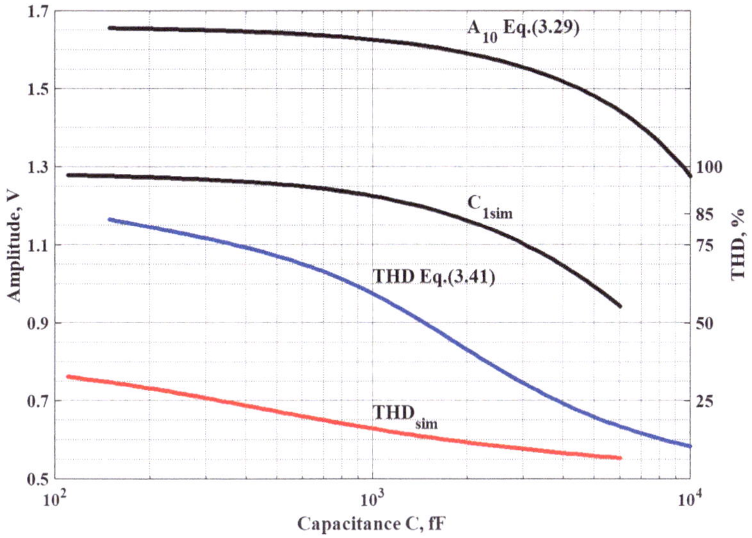

Fig. 3.9 Harmonic amplitudes and total harmonic distortion versus C and λ

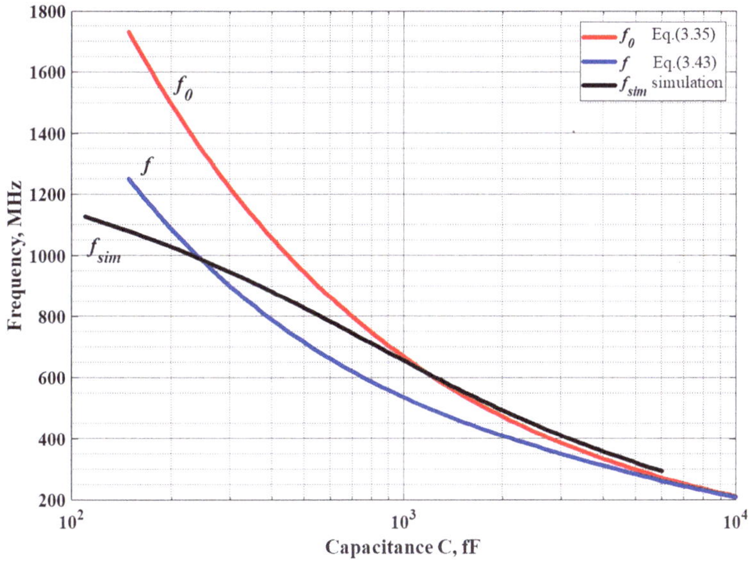

Fig. 3.10 Oscillation frequency versus capacitor

are sufficiently close, and Fig. 3.10 clearly indicates that the correction of frequency should be considered.

3.3.5 Some Observations on the van der Pol Oscillator

The comparison of the waveforms observed in simulations (Fig. 3.8) with the published results for the van der Pol model [3] shows that this model represents reasonably well the properties of the designed oscillator described in this chapter. The model gives more pessimistic results (for amplitudes, frequency reduction and THD), but this can be considered as an advantage. If the frequency control is realized by using the oscillator capacitor (which is usually the case), then reducing the capacitance to smaller values inevitably increases the harmonic distortion. The phase noise increases as well (hence, setting a certain level of THD at the start of design will also limit the oscillator phase noise), but this question is outside the scope of this book. For completeness of the results, we indicate that the phase noise of this oscillator was −133.2 dBc/Hz at 1 MHz offset from the 205 MHz carrier (with C=6 pF).

References

1. D.W. Jordan, P. Smith, *Nonlinear Ordinary Differential Equations: An Introduction for Scientists and Engineers*, 4th edn. (Oxford University Press, New York, 2007)
2. J. Groszkowski, The interdependence of frequency variation and harmonic content, and the problem of constant-frequency oscillators. Proc. Inst. Radio Eng. **21**(7), 958–981 (1933)
3. B. van der Pol, The nonlinear theory of electric oscillations. Proc. Inst. Radio Eng. **22**(9), 1051–1086 (1934)
4. H.J. Reich, A low distortion audio-frequency oscillator. Proc. Inst. Radio Eng. **25**(11), 1387–1398 (1937)
5. K. Clarke, D. Hess, *Communication Circuits: Analysis and Design* (Addison-Wesley, New York, 1971)
6. F.N.H. Robinson, Distortion in sinusoidal oscillators. Int. J. Electron. **48**(2), 137–148 (1980)
7. V. Papez, S. Papezova, Highly pure signal generators, in *9th International Conference on Telecommunication in Modern Satellite, Cable, and Broadcasting Services* (2009), pp. 526–529
8. K.F. Theodorchik, *Auto-Oscillating Systems (in Russian)* (Technical Literature Pub. House, Moscow, 1952)
9. Q. Shui-Sheng, F.I.M., On the calculation of distortions in oscillators. Int. J. Circuit Theory Appl. **17**(4), 489–494 (1989)
10. A. Buonomo, C.D. Bello, Asymptotic formulas in nearly sinusoidal nonlinear oscillators. IEEE Trans. Circuits Syst. I Fundam. Theory Appl. **43**(12), 953–963 (1996)
11. A. Buonomo, A.L. Schiavo, General nonlinear analysis of second-order oscillators. Electron. Lett. **36**(5), 396–397 (2000)
12. G. Palumbo, M. Pennisi, S. Pennisi, Analysis of harmonic distortion in the Colpitts oscillator, in *2006 13th IEEE International Conference on Electronics, Circuits and Systems* (2006), pp. 196–199
13. J. Taylor, C. Clarke, Improved harmonic analysis of RC-active phase shift oscillators, in *2008 IEEE International Symposium on Circuits and Systems* (2008), pp. 1272–1275
14. G. Palumbo, M. Pennisi, S. Pennisi, Wien-type oscillators: evaluation and optimization of harmonic distortion. IEEE Trans. Circuits Syst. II Express Briefs **55**(7), 628–632 (2008)

15. J.B. Scott, On the design of very low distortion oscillators, in *IEEE International Symposium on Circuits and Systems*, vol. 2 (1989), pp. 1419–1422
16. F. Bahmani, E. Sanchez-Sinencio, Low THD bandpass-based oscillator using multilevel hard limiter. IET Circuits Devices Syst. **1**(2), 151–160 (2007)
17. C.V. Sellathamby, M.M. Reja, L. Fu, B. Bai, E. Reid, S.H. Slupsky, I.M. Filanovsky, K. Iniewski, Noncontact wafer probe using wireless probe cards, in *IEEE International Conference on Test* (2005), pp. 6–452
18. B. Razavi, *Design of Analog CMOS Integrated Circuits* (McGraw-Hill, New York, 2000)
19. M. Urabe, Periodic solutions of van der pol's equations with large damping coefficient $\lambda = 0 \sim 10$. IRE Trans. Circuit Theory **7**(4), 382–386 (1960)
20. P. Ponzo, N. Wax, On the periodic solution of the van der pol equation. IEEE Trans. Circuit Theory **12**(1), 135–136 (1965)
21. S.-S. Qiu, I. Filanovsky, Periodic solutions of the van der pol equation with moderate values of damping coefficient. IEEE Trans. Circuits Syst. **34**(8), 913–918 (1987)
22. D.M. Binkley, *Tradeoffs and Optimization in Analog CMOS Design* (Wiley, New York, 2008)

Chapter 4
Injection Locking

4.1 Introduction

In Chap. 2, we have studied the series and parallel topologies of sinusoidal oscillators in the free-running mode, in which the oscillator input is zero. In this chapter, we study a more general case, known as the driven mode, in which the oscillator input is connected to an external periodic signal generator (a nonzero input). We use the van der Pol oscillator (VDPO) to study the coupling and derive the equations for the oscillation frequency, amplitude and phase. We consider both the series and parallel topologies of the VDPO, since we want to apply the results to the study of the coupled RC-oscillator (which is modeled by the series VDPO) and the two-integrator oscillator (modeled by the parallel VDPO).

We start by describing the synchronization of a single oscillator by an external sinusoidal source. In the following chapters, we substitute the external source by a second oscillator and study its influence on the quadrature oscillator key parameters.

Before studying coupled oscillators, the main objective of this book, we present here a description of the injection-locking mechanism, which usually is not commonly included in engineering books on oscillators. It was extensively studied by Adler [1], Kurokawa [2] and others [3, 4]. The injection locking is commonly found in frequency dividers and coupled oscillators [4]. It is also used for phase-noise improvement. In the injection-locked frequency divider (ILFD), an oscillator to be locked with a sub-harmonic of its free-running frequency, thus implementing a divide-by-two or divide-by-three circuit, as explained in [5]. Another useful application is the improvement of an oscillator phase noise by direct injection of the signal of a reference oscillator (with low phase noise). The advantage of this method is the reduction of the phase noise without requiring additional circuits and power. A comprehensive study of the above applications can be found in [4]. In this book, we focus on using the injection-locking theory to study coupled oscillators that generate quadrature signals.

J. C. F. d. A. Casaleiro et al., *Quadrature RC–Oscillators*, Analog Circuits and Signal Processing, https://doi.org/10.1007/978-3-030-00740-9_4

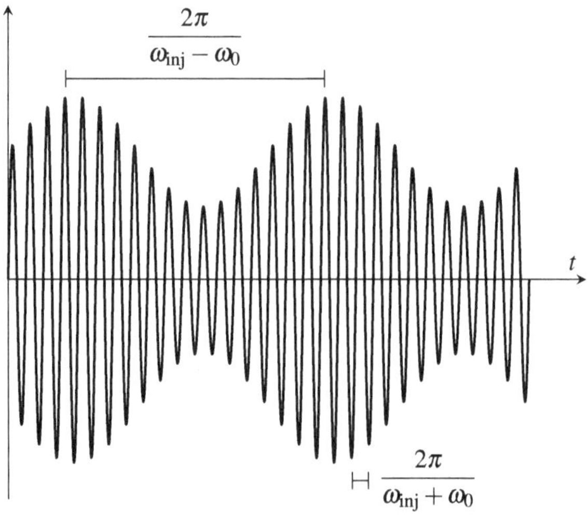

Fig. 4.1 Beating of two sinusoidal signals

Injection locking occurs when an oscillator is driven by an external periodic signal (locking signal) and the injected current, or voltage, forces the oscillator to change its frequency, synchronizing it with the locking signal. However, this synchronization occurs only if the locking frequency is within a band, known as locking range, dependent on the oscillator parameters. If the frequency is either below or above the locking range, the output will be a sinusoid (with the sum of the external and free-running frequencies) modulated in amplitude by a low-frequency sinusoid (the difference between the two frequencies), as shown in Fig. 4.1. This phenomenon is called "beating".

The locking range is an important parameter for coupled oscillators, because practical oscillators have mismatches and their oscillation frequencies may be different from their calculated values. The mismatches and the parasitic elements should not move the oscillation frequencies far outside the locking range, otherwise the locking between the two oscillators does not occur, leading to an undesired beat signal, as shown in Fig. 4.1. As will be explained below, the steady-state amplitude of the oscillator in the locked condition will depend on the phase and amplitude of the locking signal.

The phase difference, between the locking signal and the oscillator output signal, depends on the frequencies and the amplitudes of the free-running and external signals.

4.2 Parallel VDPO

In this section, we study the injection lock in the parallel VDPO. The results obtained here will be relevant to the study of the two-integrator oscillator, in Chap. 7, which consists of two integration stages coupled by transconductance amplifiers. The two-integrator oscillator can be modeled by two injection-locked stages, in

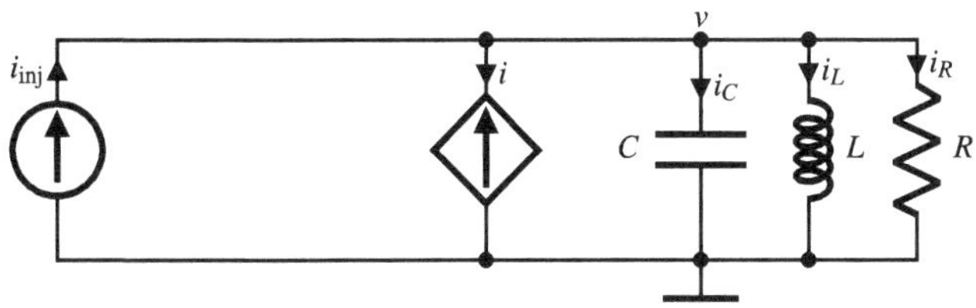

Fig. 4.2 Injection lock of a parallel VDPO

which the output of a stage will provide the injection current to the other stage. We will analyse the parallel VDPO with an external sinusoidal current source i_{inj} (locking signal) in parallel, as shown in Fig. 4.2. Applying the Kirchhoff's current law (KCL), we obtain

$$i_C + i_R + i + i_L = i_{inj}, \tag{4.1}$$

where $i = K_0 v - K_2 v^3$.

This gives us

$$C\frac{dv}{dt} + \frac{1}{R}v - K_0 v + K_2 v^3 + \frac{1}{L}\int v dt = i_{inj}, \tag{4.2}$$

Dividing both sides of Eq. (4.2) by C and differentiating, one obtains

$$\frac{d^2 v}{dt^2} - 2\left(\delta_0 - \delta_2 v^2\right)\frac{dv}{dt} + \omega_0^2 v = \frac{\omega_0 R}{Q}\frac{di_{inj}}{dt}, \tag{4.3}$$

where

$$\delta_0 = \frac{\omega_0}{2Q}\left(K_0 R - 1\right), \qquad \delta_2 = \frac{3\omega_0}{2Q}K_2 R,$$

and Q is the quality factor of the parallel RLC circuit. We call Eq. (4.3) the equation of driven VDPO because the right-hand side is nonzero. In mathematical terminology, this is a non-homogeneous differential equation. From differential equations theory, it is known that the general solution of a linear differential equation is the solution of the homogeneous equation, v_H, plus the particular solution of the non-homogeneous equation, v_P:

$$v(t) = v_H(t) + v_P(t). \tag{4.4}$$

Although the left-hand side of Eq. (4.3) is not linear, it can be approximated by a linear equation if the amplitude variation is small making the dv/dt term very small [6], which is true near the steady state. Hence, assuming that the system is near steady state and knowing that the solution is sinusoidal, Eq. (4.4) can be written as

$$v(t) = V \sin(\omega t - \phi) = (V_H + V_P)\sin[\omega t - (\phi_H + \phi_P)],$$

where V_H and ϕ_H are the amplitude and phase of the homogeneous solution, and V_P and ϕ_P are the amplitude and phase of the particular solution. The amplitude and phase derivatives of the homogeneous solution are

$$\begin{cases} \dfrac{\mathrm{d}V_H}{\mathrm{d}t} = \delta_0 V_H - \dfrac{1}{4}\delta_2 V_H^3 & (4.5a) \\ \dfrac{\mathrm{d}\phi_H}{\mathrm{d}t} = 0. & (4.5b) \end{cases}$$

as shown in Chap. 3, Eq. (3.14).

Now, we assume a locking signal of the form:

$$i_{\text{inj}} = I_{\text{inj}} \sin\left(\omega_{\text{inj}} t - \phi_{\text{inj}}\right). \tag{4.6}$$

The differentiation of Eq. (4.6) gives

$$\frac{\mathrm{d}i_{\text{inj}}}{\mathrm{d}t} = \frac{\mathrm{d}I_{\text{inj}}}{\mathrm{d}t} \sin\left(\omega_{\text{inj}} t - \phi_{\text{inj}}\right) + I_{\text{inj}}\left(\omega_{\text{inj}} - \frac{\mathrm{d}\phi_{\text{inj}}}{\mathrm{d}t}\right)\cos\left(\omega_{\text{inj}} t - \phi_{\text{inj}}\right),$$

which for $d\phi_{inj}/dt \approx 0$ and for a slow varying amplitude (i.e. $dI_{inj}/dt \approx 0$) is simplified to:

$$\frac{\mathrm{d}i_{\text{inj}}}{\mathrm{d}t} \approx I_{\text{inj}}\omega_{\text{inj}} \cos\left(\omega_{\text{inj}} t - \phi_{\text{inj}}\right), \tag{4.7}$$

Now, we use the harmonic balance method, which consists of matching the left- and right-hand sides of the differential equation (4.3). Note that the left-hand side of Eq. 4.3 can be expressed in $\sin \omega t$ and $\cos \omega t$ terms. To match with the right-hand side, we write Eq. (4.7) in terms of $\sin \omega t$ and $\cos \omega t$ using the trigonometric relationship $\cos(\alpha + \beta) = \cos\alpha \cos\beta - \sin\alpha \sin\beta$, resulting in:

$$\frac{\mathrm{d}i_{\text{inj}}}{\mathrm{d}t} = I_{\text{inj}}\omega_{\text{inj}}\left[\cos\left(\Omega t - \phi_{\text{inj}}\right)\cos\omega t - \sin\left(\Omega t - \phi_{\text{inj}}\right)\sin\omega t\right], \tag{4.8}$$

where $\Omega = (\omega_{\text{inj}} - \omega)$ is the difference between the frequency of the injected signal and the oscillator's free-running frequency (it is zero when there is locking). Substituting Eq. (4.8) into Eq. (4.3) and simplifying the second term of Eq. (4.3), we obtain the amplitude and phase derivatives of the particular solution:

$$\begin{cases} \dfrac{\mathrm{d}V_P}{\mathrm{d}t} = \dfrac{\omega_0}{Q} \cdot \dfrac{\omega_{\text{inj}}}{2\omega} \cdot RI_{\text{inj}} \cos\left[\left(\omega_{\text{inj}} - \omega\right)t + \phi\right] & (4.9a) \\ \dfrac{\mathrm{d}\phi_P}{\mathrm{d}t} = \dfrac{\omega^2 - \omega_0^2}{2\omega} - \dfrac{\omega_0}{Q} \cdot \dfrac{\omega_{\text{inj}}}{2\omega} \cdot \dfrac{RI_{\text{inj}}}{V} \sin\left[\left(\omega_{\text{inj}} - \omega\right)t + \phi\right], & (4.9b) \end{cases}$$

where ϕ is the phase difference between the oscillator output and the locking signal, and I_{inj} is the amplitude of the locking signal. Adding Eq. (4.5) and Eq. (4.9), we obtain the amplitude and phase derivatives of the general solution:

$$\begin{cases} \dfrac{\mathrm{d}V}{\mathrm{d}t} = \left(\delta_0 V - \dfrac{1}{4}\delta_2 V^3\right) + \dfrac{\omega_0\omega_{\text{inj}}}{2Q\omega} R I_{\text{inj}} \cos\left[\left(\omega_{\text{inj}} - \omega\right) t + \phi\right] & \text{(4.10a)} \\[2ex] \dfrac{\mathrm{d}\phi}{\mathrm{d}t} = \dfrac{\omega^2 - \omega_0^2}{2\omega} - \dfrac{\omega_0\omega_{\text{inj}} R I_{\text{inj}}}{2Q\omega V} \sin\left[\left(\omega_{\text{inj}} - \omega\right) t + \phi\right]. & \text{(4.10b)} \end{cases}$$

where V is the amplitude and ω_0 is the oscillator natural frequency of oscillation. These equations are non-autonomous, since their right-hand side is not constant and varies with time t. To have a steady state with constant amplitude and phase, these equations should be autonomous, i.e. their right-side should be constant. This is possible only if the oscillator is locked to the external signal (i.e. $\omega = \omega_{\text{inj}}$). Otherwise, the amplitude and phase vary with t, generating beats.

Assuming that the oscillator is locked, $\omega = \omega_{\text{inj}}$, Eq. (4.10) becomes

$$\begin{cases} \dfrac{\mathrm{d}V}{\mathrm{d}t} = \left(\delta_0 V - \dfrac{1}{4}\delta_2 V^3\right) + \dfrac{\omega_0}{2Q} R I_{\text{inj}} \cos(\phi) & \text{(4.11a)} \\[2ex] \dfrac{\mathrm{d}\phi}{\mathrm{d}t} = \dfrac{\omega_{\text{inj}}^2 - \omega_0^2}{2\omega_{\text{inj}}} - \dfrac{\omega_0}{2Q} \cdot \dfrac{I_{\text{inj}}}{I} \sin(\phi). & \text{(4.11b)} \end{cases}$$

We analyse first the phase derivative given by Eq. (4.11b) and illustrated in Fig. 4.3. In Fig. 4.3, the locking region (in which the oscillator frequency equals

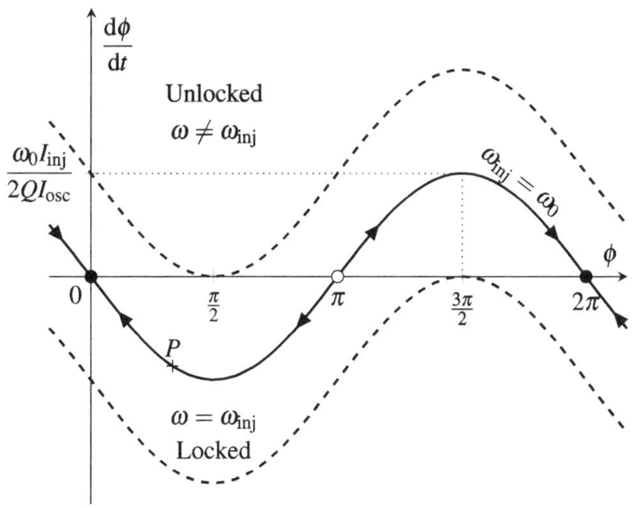

Fig. 4.3 Phase curve of the injection-lock parallel VDPO

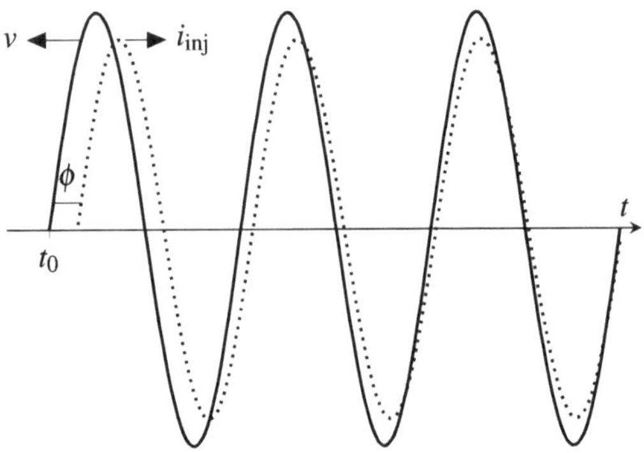

Fig. 4.4 Time solution of path P

the locking frequency) is represented in white and the unlocking regions (where $\omega \neq \omega_{inj}$) are shaded. The dashed curves represent the boundary between two regions. To better understand the circuit behaviour, we consider the particular case of a locking signal with the frequency of the free-running oscillator, $\omega_{inj} = \omega_0$ (represented in the figure by the solid curve).

From Eq. (4.11b), we find two equilibrium points: a stable equilibrium point at $\phi = 0$ and an unstable at $\phi = \pi$. Hence, in steady state, the locking signal and the oscillator output will be in-phase.

Let us consider the signals of Fig. 4.4 with a phase difference $\phi = \pi/3$, which corresponds to point P in the phase curve of Fig. 4.3. At this point, the phase derivative is negative, meaning that the oscillator phase will be forced to decrease until both signals are in-phase. Hence, as time passes, the phase difference is reduced, as shown in Fig. 4.4. If we consider a phase difference higher than $\phi = \pi$, the derivative is positive, meaning that the phase will increase until it reaches $\phi = 2\pi$.

From Fig. 4.3, we can see that other curves that are offset versions of this particular case (that will be within the white region) are possible. For $\omega_{inj} > \omega_0$, the curve is shifted upwards and the stable equilibrium point shifts to the right (increasing the phase difference). For $\omega_{inj} < \omega_0$, the curve is shifted downwards and the stable equilibrium point shifts to the left (decreasing the phase difference).

Let us consider a frequency sweep of the locking signal. We start with a frequency equal to ω_0 having a zero phase difference. As the frequency increases so does the phase difference, until the high boundary is reached (higher dashed curve in Fig. 4.3). At this point, a single equilibrium point exists (at $\phi = \pi/2$), but it is unstable, meaning that the oscillator cannot follow further the locking signal and enters the unlocking region. This frequency is the upper limit of the locking range, $\Delta\omega$. In the opposite direction, if we decrease the frequency until the bottom limit is reached, the same happens (the only difference is that the unstable equilibrium point is at $\phi = 3\pi/2 = -\pi/2$). From Eq. (4.11b), we can analytically determine the locking range by equating the left-hand side to zero and solving to ω_{inj} for the two boundary cases. Hence, for $\phi = -\pi/2$,

$$\omega_{\text{inj.min}}^2 + \frac{\omega_0}{Q} \cdot \frac{I_{\text{inj}}}{I} \omega_{\text{inj.min}} - \omega_0^2 = 0, \tag{4.12}$$

so that the positive root is

$$\omega_{\text{inj.min}} = -\frac{\omega_0}{2Q}\frac{I_{\text{inj}}}{I} + \frac{1}{2}\sqrt{\left(\frac{1}{2Q}\frac{I_{\text{inj}}}{I}\right)^2 \omega_0^2 + 4\omega_0^2}. \tag{4.13}$$

For $\omega_{\text{inj}} > \omega_0$ and $\phi = \pi/2$,

$$\omega_{\text{inj.max}}^2 - \frac{\omega_0}{Q} \cdot \frac{I_{\text{inj}}}{I} \omega_{\text{inj.max}} - \omega_0^2 = 0, \tag{4.14}$$

and the positive root is

$$\omega_{\text{inj.max}} = \frac{\omega_0}{2Q}\frac{I_{\text{inj}}}{I} + \frac{1}{2}\sqrt{\left(\frac{1}{2Q}\frac{I_{\text{inj}}}{I}\right)^2 \omega_0^2 + 4\omega_0^2}. \tag{4.15}$$

The difference between the maximum Eq. (4.15) and minimum Eq. (4.13) frequencies defines the locking range:

$$\Delta\omega = \omega_{\text{inj.max}} - \omega_{\text{inj.min}} = \frac{\omega_0}{Q}\frac{I_{\text{inj}}}{I}. \tag{4.16}$$

By taking again Eq. (4.11b), equating the left-hand side to zero and solving the equation for ϕ, we get

$$\phi = \arcsin\left(\frac{Q}{\omega_0}\frac{I}{I_{\text{inj}}}\frac{\omega_{\text{inj}}^2 - \omega_0^2}{\omega_{\text{inj}}}\right), \tag{4.17}$$

which relates the phase difference with the frequency of the locking signal, as shown in Fig. 4.5. From Fig. 4.5, we can see that the phase difference decreases with the increase of the locking frequency. Near the free-running frequency, ω_0, the characteristic is almost linear with the slope of $Q/\omega_0 I_{\text{inj}}$. It should be noted that the graph in Fig. 4.5 is not defined for frequencies outside the locking range, since the oscillator is not locked and the phase is not constant.

Let us analyse now the amplitude derivative given by Eq. (4.10a) and illustrated in Fig. 4.6, where the shaded areas correspond to the states where the oscillator is unlocked. The white area corresponds to the states where the oscillator is locked and described by Eq. (4.11a). We rewrite it here for convenience:

$$\frac{dV}{dt} = \left(\delta_0 V - \frac{1}{4}\delta_2 V^3\right) + \frac{\omega_0}{2Q} R I_{\text{inj}} \cos\phi. \tag{4.18}$$

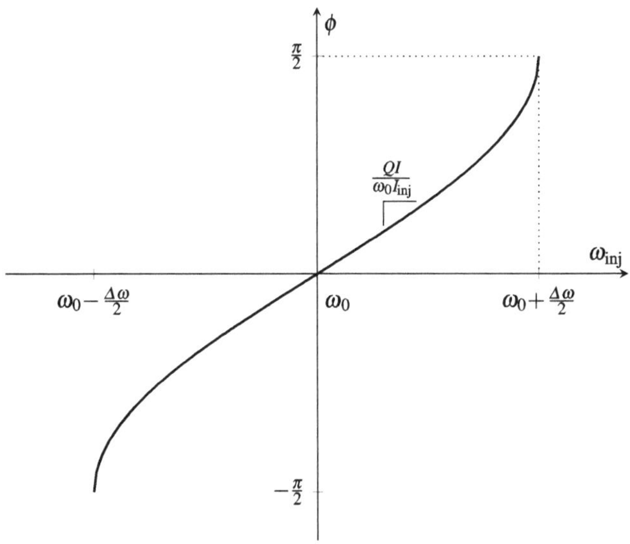

Fig. 4.5 Injection-lock phase curve

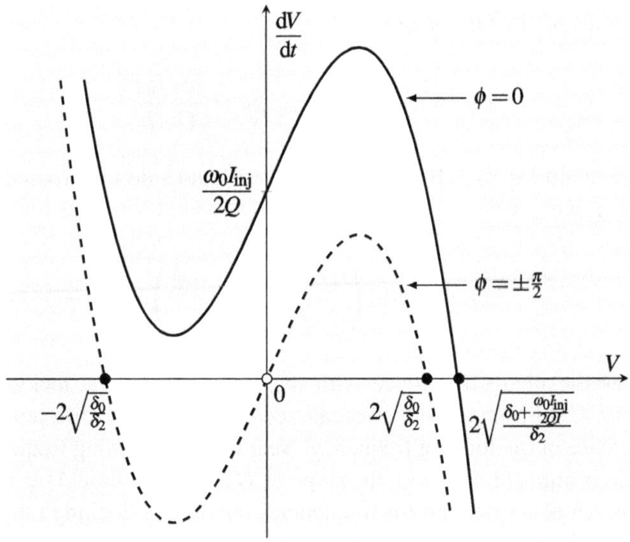

Fig. 4.6 Amplitude phase curve of the injection locking

The dashed curve represents the boundary between the locking and unlocking areas. Notice that at the boundary the oscillator is unlocked, although it seems to be in a steady-state solution, since it has stable equilibrium points. The reason for that is that this boundary occurs for a phase difference of $\phi = \pi/2$, or $\phi = -\pi/2$, which are both unstable in the phase curve, as shown in Fig. 4.3. From Eq. (4.18), we can see that the other cases, within the white area, are the offset versions of the boundary. The upper limit occurs when the locking and the free-running frequencies are equal forcing the oscillator to be in-phase with the locking signal. Note that the stable equilibrium points are shifted to the right as the offset increases, which is equivalent to an increase of the amplitude. Solving Eq. (4.18) for steady state (i.e. $\mathrm{d}V/\mathrm{d}t = 0$) and substituting δ_0 and δ_2 by the equations in terms of circuit parameters results in:

$$V \approx 2\sqrt{\frac{\left(K_0 + \frac{I_{\text{inj}}}{2}\cos\phi\right)R - 1}{3K_2R}}, \qquad \text{for } V > 0, \tag{4.19}$$

which relates the amplitude with the phase difference, as shown in Fig. 4.6. The curve in Fig. 4.6 is an approximate representation, since the relationship between the phase and the frequency is not linear, especially near the boundary, as can be seen from Fig. 4.6.

Hence, if the frequency of the external signal is within the locking range, the oscillator locks in frequency, i.e. oscillates with the frequency of external signal. Although the frequency locks, the phase and amplitude depend on the locking frequency. From Eq. (4.17), we conclude that for frequencies near ω_0, the phase is proportional to Q and to the ratio between the oscillator current, I, and the injected current, I_{inj}:

$$\phi \propto \left(\frac{Q}{\omega_0} \cdot \frac{I}{I_{\text{inj}}}\right)(\omega_{\text{inj}} - \omega_0). \tag{4.20}$$

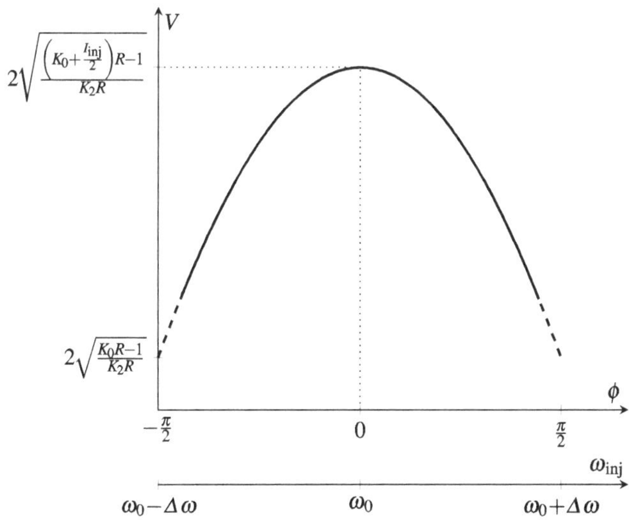

Fig. 4.7 Amplitude as a function of the oscillation frequency

From Eq. (4.19), we conclude that the amplitude depends on the phase (which in turn depends on the locking frequency) and on the injection signal amplitude, I_{inj}, as shown in Fig. 4.7.

4.3 Series VDPO

In this section, we analyse the injection locking of a series VDPO, which is relevant because the RC-oscillators are best modeled by the series VDPO. The results obtained here will be useful for the analysis of the active and passive coupling of RC-oscillators in Chaps. 5 and 6. We first analyse the injection locking with a single external source (similar to the previous section), which is suitable for the active coupling. After that, we analyse the double injection, which is best suited for the analysis of oscillators with passive coupling. The analysis of the passive coupling is more cumbersome, because the current injected into an oscillator is forced by another oscillator, meaning that each oscillator will have two injection currents, one coming in and the other going out.

4.3.1 Single External Source

Let us analyse the series VDPO with a single external current source (locking signal) in parallel with the capacitor, as shown in Fig. 4.8, where $f(i) = (-K_0 i + K_2 i^3)$ represents the VDP nonlinearity and R the loss of the inductor. Applying the Kirchhoff's voltage law (KVL), we obtain

$$v_L - v_f + v_R - v_C = 0.$$

Substituting the currents and voltages by the elements' equations, we get

$$L\frac{\mathrm{d}i}{\mathrm{d}t} + Ri + K_0 i - K_2 i^3 + \frac{1}{C}\int i\mathrm{d}t = \frac{1}{C}\int i_{\mathrm{inj}}\mathrm{d}t. \tag{4.21}$$

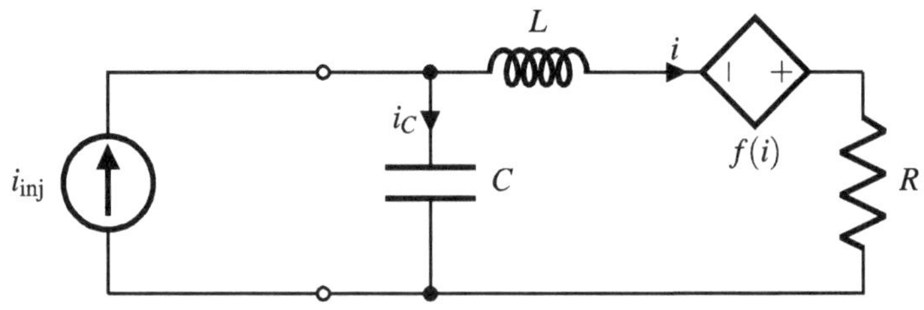

Fig. 4.8 Series van der Pol oscillator driven by external source

Dividing by L and differentiating both sides of this equation results in:

$$\frac{\mathrm{d}^2 i}{\mathrm{d}t^2} - 2\left(\delta_0 - \delta_2 i^2\right)\frac{\mathrm{d}i}{\mathrm{d}t} + \omega_0^2 i = \omega_0^2 i_{\mathrm{inj}}, \tag{4.22}$$

where

$$\delta_0 = -\frac{R + K_0}{2L}; \qquad \delta_2 = -3\frac{K_2}{2L}.$$

Comparing Eq. (4.3) with Eq. (4.22) shows that the two systems are not fully dual, since the right-hand side of Eq. (4.22) depends directly on the locking signal and not on its derivative. From Eq. (4.22), taking, the forced solution of the form:

$$i(t) = I \sin(\omega t - \phi),$$

and a locking signal of the form given by Eq. (4.6). Then, assuming that the oscillator is locked ($\omega = \omega_{\mathrm{inj}}$), we obtain the amplitude and phase derivatives:

$$\begin{cases} \dfrac{\mathrm{d}I}{\mathrm{d}t} = \left(\delta_0 I - \dfrac{1}{4}\delta_2 I^3\right) + \dfrac{\omega_0^2}{2\omega_{\mathrm{inj}}} I_{\mathrm{inj}} \sin(\phi) & (4.23\mathrm{a}) \\ \dfrac{\mathrm{d}\phi}{\mathrm{d}t} = \dfrac{\omega_{\mathrm{inj}}^2 - \omega_0^2}{2\omega_{\mathrm{inj}}} + \dfrac{\omega_0^2}{2\omega_{\mathrm{inj}}} \cdot \dfrac{I_{\mathrm{inj}}}{I} \cos\phi. & (4.23\mathrm{b}) \end{cases}$$

One can conclude from Eq. (4.23) that the stable equilibrium point for the phase difference is shifted by $\pi/2$, in comparison with the driven parallel VDPO, as shown in Fig. 4.9. Hence, a locking signal with the frequency near the free-running frequency of the oscillator, ω_0, forces a phase difference of $\pi/2$ between the signals, as shown by the injection-lock phase curve shown in Fig. 4.10.

Moreover, the locking range is given by:

$$\Delta\omega \approx 2\omega_0 \frac{I_{\mathrm{inj}}}{I}.$$

which is higher than that of the driven parallel VDPO.

4.3.2 *Two External Sources*

Let us now analyse the series VDPO with two external current sources (locking signals), as shown in Fig. 4.11. This type of injection locking and the study presented

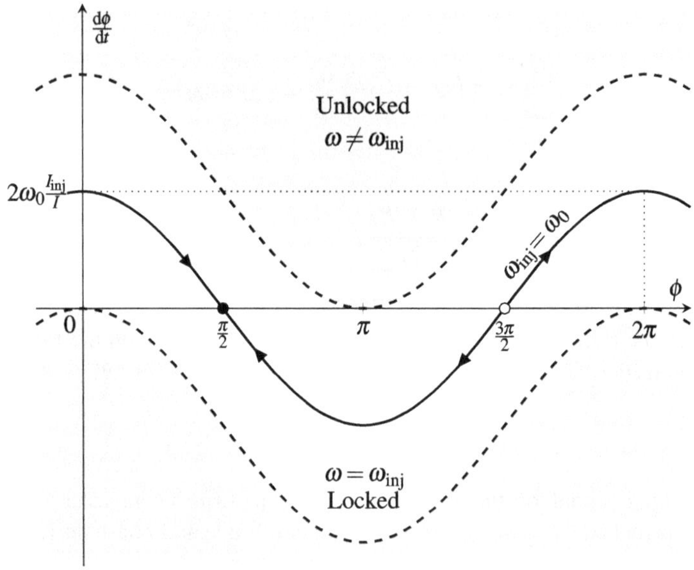

Fig. 4.9 Phase curve of the series VDPO

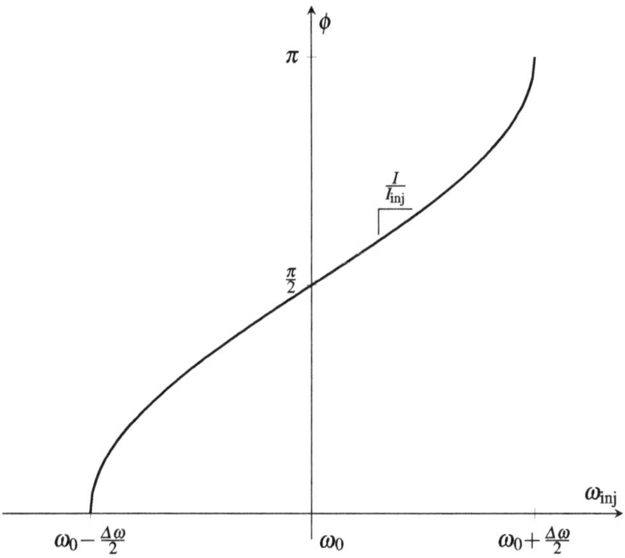

Fig. 4.10 Phase as a function of the frequency of the external signal

here is useful to understand the coupling using passive networks, described in Chap. 6. Usually, passive coupling networks are reciprocal, which means that the output current of an oscillator is injected into the second oscillator and the other way around. This behaviour is well modeled by the double injection locking, which

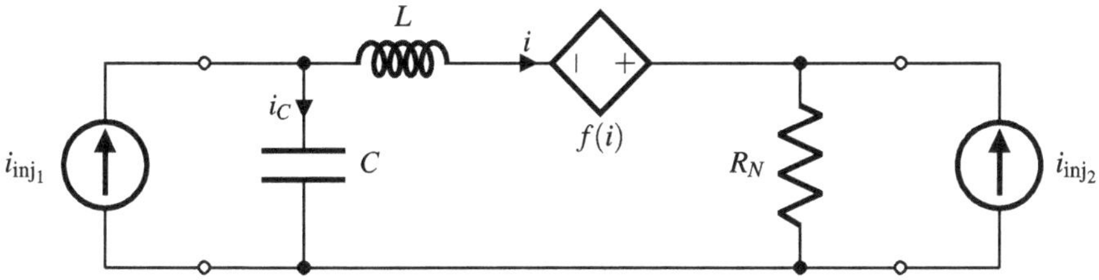

Fig. 4.11 Driven series van der Pol oscillator with double injection

is the main reason for presenting the study that follows. Applying the KCL and KVL to the circuit Fig. 4.11, we obtain

$$\begin{cases} i_C + i = i_{inj_1} & (4.24a) \\ i_{R_N} = i + i_{inj_2} & (4.24b) \\ v_L + f(i) + v_{R_N} - v_C = 0 & (4.24c) \end{cases}$$

Substituting into Eq. (4.24) the currents and voltages by their equations for the network elements, one obtains

$$L\frac{di}{dt} + R_N i + K_0 i - K_2 i^3 + \frac{1}{C}\int i dt = \frac{1}{C}\int i_{inj_1} dt + R_N i_{inj_2}.$$

Dividing by L and differentiating both sides of this equation gives the result:

$$\frac{d^2 i}{dt^2} - 2\left(\delta_0 - \delta_2 i^2\right)\frac{di}{dt} + \omega_0^2 i = \omega_0^2 i_{inj_1} + \frac{R_N}{L}\frac{di_{inj_2}}{dt}, \tag{4.25}$$

where the VDP parameters are given by:

$$\delta_0 = -\frac{R_N + K_0}{2L}; \qquad \delta_2 = -3\frac{K_2}{2L}.$$

From Eq. (4.25), one can see that the solution depends on both locking signals, i_{inj_1} and i_{inj_2}, as expected. However, due to the many possibilities, we have to simplify the problem by assuming that both signals have the same frequency and phase, and the oscillator is locked to both signals of the same frequency (i.e. $\omega_{inj1} = \omega_{inj2} = \omega_{inj}$). With both locking signals of the form given by Eq. (4.6) applied to the locked oscillator, we obtain the amplitude and phase derivatives:

$$\begin{cases} \dfrac{dI}{dt} = \left(\delta_0 I - \dfrac{1}{4}\delta_2 I^3\right) + \dfrac{\omega_0^2}{2\omega_{inj}} I_{inj1}\sin(\phi) + \omega\dfrac{R_N}{L} I_{inj2}\cos(\phi) & (4.26a) \\ \dfrac{d\phi}{dt} = \dfrac{\omega^2 - \omega_0^2}{2\omega} + \dfrac{\omega_0^2}{2\omega}\cdot\dfrac{I_{inj1}}{I}\cos(\phi) - \dfrac{R_N}{2\omega L}\cdot\dfrac{I_{inj2}}{I}\sin(\phi). & (4.26b) \end{cases}$$

From Eq. (4.26), we can see that for both amplitude and phase, the effect of two injected signals can be combined into one single injection signal. However, the result and conclusions that we can obtain here allow better interpretation if we apply them to some practical oscillators, after we have studied the passive coupling, using the output signals of each oscillator.

4.4 Conclusion

From the analysis of three injection-locking topologies, we can conclude that the driven VDPO locks to the frequency of the locking signal within a limited frequency range (locking range). The phase and amplitude are adjusted according to the attenuation and phase imposed by the oscillator's resonant tank at the locked frequency. For example, the parallel topology imposes a relative phase between the oscillator and the locking signal in the range of $-\pi/2$ to $\pi/2$. The series topology relative phase is within the range of 0 to π. The oscillation amplitude has its peak at the resonant frequency of the tank, ω_0, but away from the resonant frequency the amplitude drops. Moreover, the series topology has a higher locking range than the parallel one, and the relationship between the phase and frequency has a higher slope.

It is worth mentioning that, although we analysed the injection locking assuming sinusoidal sources, the signals of other shapes can be also used. The only requirement is that the locking signal has to be periodic. Knowing, from the Fourier series, that any periodic signal can be represented by a sum of sines and cosines, the injection-locking analysis can be generalized to any periodic signal if we use the Fourier series of the locking signal in the right-hand side of Eq. (4.3). Moreover, if the resonator has enough selectivity we can assume that the high-order harmonics are strongly attenuated and locking is reduced to that of the fundamental frequency (first harmonic only). This means that for a highly-selective resonator, the theory presented in this section can be used. We only need to calculate one coefficient of the Fourier series and substitute in I_{inj_1}. However, for a low-selectivity resonator, like those in the RC-oscillators considered in the next chapters, the high-order harmonics are not attenuated sufficiently, so we cannot neglect these harmonics.

References

1. R. Adler, A study of locking phenomena in oscillators. Proc. Inst. Radio Eng. **34**(6), 351–357 (1946)
2. K. Kurokawa, *An Introduction to the Theory of Microwave Circuits* (Academic, New York, 1969)
3. L. Paciorek, Injection locking of oscillators. Proc. IEEE **53**(11), 1723–1727 (1965)
4. B. Razavi, A study of injection locking and pulling in oscillators. IEEE J. Solid State Circuits **39**(9), 1415–1424 (2004)

5. A. Buonomo, A. Lo Schiavo, M. Awan, M. Asghar, M. Kennedy, A CMOS injection-locked frequency divider optimized for divide-by-two and divide-by-three operation. IEEE Trans. Circuits Syst. I **60**(12), 3126–3135 (2013)
6. D.W. Jordan, P. Smith, *Nonlinear Ordinary Differential Equations: An Introduction for Scientists and Engineers*, 4th edn. (Oxford University Press, New York, 2007)

Chapter 5
Active Coupling *RC*-Oscillator

5.1 Introduction

In this chapter, we study the actively cross-coupled *RC*-oscillator, which consists of two *RC*-oscillators coupled by transconductance amplifiers. The single and cross-coupled relaxation oscillators were extensively studied in [1], where a comprehensive analysis of the cross-coupled *RC*-oscillator in relaxation regime can be found. Here, we study the cross-coupled *RC*-oscillator in nearly sinusoidal regime with low coupling factor. For a strong coupling, amplifiers are approximated by hard limiters, injecting a square wave current into the other oscillator [2]. In oscillators with high-quality factor-resonant tanks (such as the *LC*-oscillators), the high-order harmonics are filtered out and the injected signal is reduced to the first harmonic of the Fourier series. However, to ensure that low-quality factor oscillators (as the *RC*-oscillator) are in the sinusoidal regime, a low coupling factor is necessary; hence, the high-order harmonics are not sufficiently attenuated. The assumption of low coupling strength makes the analysis more cumbersome.

We first review the single *RC*-oscillator using the approximation proposed in [3], in which the *RC*-oscillator working in the nearly sinusoidal regime can be modeled by an *RLC* circuit. We approximate it to the VDPO as in [4], to account for the amplitude control mechanism. Next, we substitute each *RC*-oscillator by a VDPO and the transconductance amplifiers by voltage-controlled current sources, reducing each oscillator to the series injected locking circuit of Chap. 4. Afterwards, we derive the quadrature oscillator key parameters: amplitude, frequency and quadrature error. The parameters are obtained for steady state assuming no mismatches. This derivation is followed by the stability analysis of the steady-state solution. Then, assuming mismatches and steady state, we derive the amplitude- and phase-error equations. In the last section, we draw the conclusions.

J. C. F. d. A. Casaleiro et al., *Quadrature RC–Oscillators*, Analog Circuits and Signal Processing, https://doi.org/10.1007/978-3-030-00740-9_5

5.2 Single RC-Oscillator

In this section, the single RC-oscillator, as shown in Fig. 5.1a, is analysed assuming ideal current sources and mismatches only in the transistors' transconductances. The circuit in analysed in the nearly sinusoidal regime and it is shown that it can be approximated by the series VDPO. The small-signal equivalent circuit of the single oscillator is shown in Fig. 5.1b. Here, capacitance C_d is the capacitance of the transistors, and C_p (which is not shown in the circuit) represents other parasitic capacitances lumped together. We approximate C_d as:

$$C_d \approx 2C_{gd} + \frac{1}{2}C_{gs} + C_p. \tag{5.1}$$

It should be noted that, although the capacitance C_d does not appear explicity in Fig. 5.1a, it is present below in the equations for L_e and R_N.

The parameters G_1 and G_2 represent the signal-dependent transconductances of M_1 and M_2, respectively. It should be noted that G_1 and G_2 are signal dependent. Their approximation for the following time-domain analysis was considered in Chap. 3.

Applying the Kirchhoff's voltage and current laws to the small-signal circuit in Fig. 5.1b, we obtain

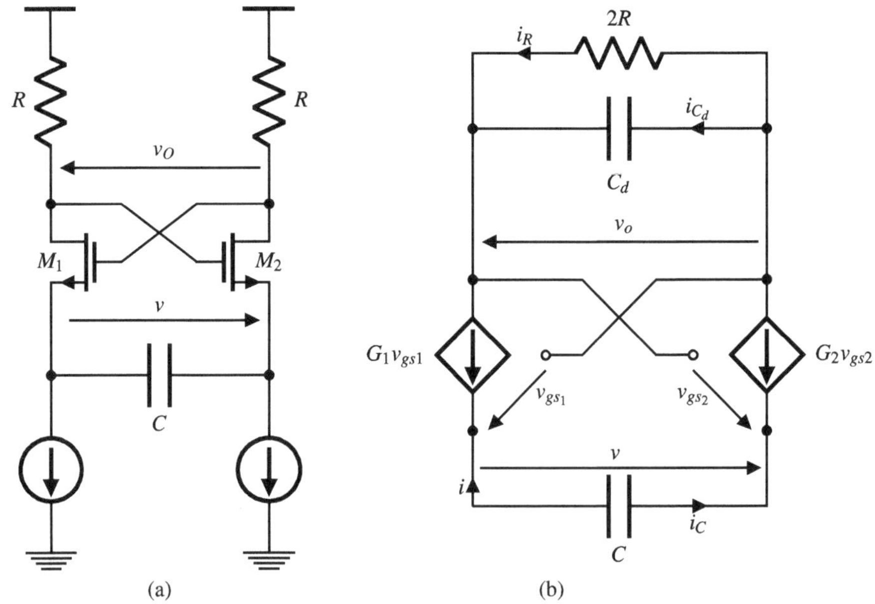

Fig. 5.1 Single RC-oscillator (**a**) circuit and (**b**) small-signal equivalent circuit

$$\begin{cases} i = -G_1 v_{gs1} & (5.2a) \\ i = G_2 v_{gs2} & (5.2b) \\ i = -i_{Cd} - i_R = -C_d \dfrac{\mathrm{d}v_o}{\mathrm{d}t} - \dfrac{v_o}{2R} & (5.2c) \\ v_o = v_{gs1} - v_{gs2} + v & (5.2d) \\ i = -i_c = -C \dfrac{\mathrm{d}v}{\mathrm{d}t} & (5.2e) \end{cases}$$

Substituting Eq. (5.2a) and Eq. (5.2b) into Eq. (5.2d), one obtains

$$v_o = -\frac{i}{G_1} - \frac{i}{G_2} + v \tag{5.3}$$

Substituting this result to Eq. (5.2c) and solving with respect to v gives

$$v \approx \left(\frac{G_1 + G_2}{G_1 \cdot G_2}\right) i - 2R\left(1 - \frac{C_d}{C}\right) i + 2RC_d\left(\frac{G_1 + G_2}{G_1 \cdot G_2}\right)\frac{\mathrm{d}i}{\mathrm{d}t}, \tag{5.4}$$

The first term on the right-hand side of Eq. (5.4) represents a nonlinear resistance, the second term a negative resistance that compensates the loss in the nonlinear resistance and the last term represents a nonlinear inductance. This last term on the right-hand side of Eq. (5.4) is an approximation [3, 4]. If the incremental current, i, is small, the nonlinear terms of the signal-dependent transconductances, G_1 and G_2, are small in comparison with the linear term. With this assumption, we can assume that:

$$\frac{\mathrm{d}}{\mathrm{d}t}\left[\frac{G_1 + G_2}{G_1 \cdot G_2} i\right] \approx \left(\frac{G_1 + G_2}{G_1 \cdot G_2}\right)\frac{\mathrm{d}i}{\mathrm{d}t}.$$

Thus, the single *RC*-oscillator can be substituted by the equivalent circuit shown in Fig. 5.2 [3, 4]. In this circuit, the equivalent inductance is given by:

$$L_e \approx 4RC_d g_{m0}{}^{-1}, \tag{5.5}$$

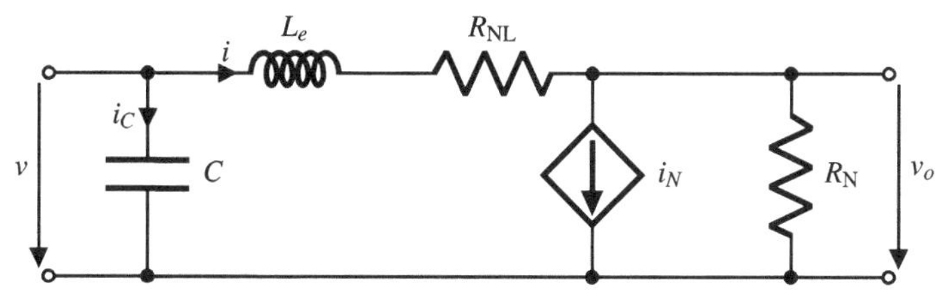

Fig. 5.2 Equivalent circuit of a single *RC*-oscillator

where g_{m0} is the transconductance of $M_{1,2}$. The positive resistance, generated by the cross-coupled pair $M_{1,2}$, is given by:

$$R_{\mathrm{NL}} = 2{g_{m0}}^{-1}, \tag{5.6}$$

the negative resistance is:

$$R_{\mathrm{N}} = -2R\left(1 - \frac{C_d}{C}\right), \tag{5.7}$$

and the dependent current source, related to the nonlinear resistance, is:

$$i_N = K^2 i^3, \tag{5.8}$$

where K is a parameter that depends on the transistor working region (see Chap. 3).

Substituting Eq. (5.4) into Eq. (5.2e) and rearranging the terms, we obtain the differential equation of the circuit in Fig. 5.2 as:

$$\frac{\mathrm{d}^2 i}{\mathrm{d}t^2} + \left(\frac{R_{\mathrm{NL}} + R_{\mathrm{N}}}{L_e}\right)\frac{\mathrm{d}i}{\mathrm{d}t} - \left(\frac{R_{\mathrm{N}}}{L_e}\right)\frac{\mathrm{d}i_N}{\mathrm{d}t} + \frac{1}{L_e C} i \approx 0. \tag{5.9}$$

Substituting Eq. (5.5), Eq. (5.6), Eq. (5.7) and Eq. (5.8) into Eq. (5.9), we obtain

$$\frac{\mathrm{d}^2 i}{\mathrm{d}t^2} + 2\frac{1 - Rg_{m0}\left(1 - \frac{C_d}{C}\right) + 3Rg_{m0}\left(1 - \frac{C_d}{C}\right)K^2 i^2}{4RC_d}\frac{\mathrm{d}i}{\mathrm{d}t} + \frac{g_{m0}}{4RC_d C} i = 0. \tag{5.10}$$

The differential equation (5.10) can be written in the form:

$$\frac{\mathrm{d}^2 i}{\mathrm{d}t^2} - 2\left(\delta_0 - \delta_2 i^2\right)\frac{\mathrm{d}i}{\mathrm{d}t} + \omega_0^2 i = 0. \tag{5.11}$$

This is the equation of the VDPO with the parameters:

$$\delta_0 = \frac{Rg_{m0}\left(1 - \frac{C_d}{C}\right) - 1}{4RC_d}; \qquad \delta_2 = \frac{3Rg_{m0}\left(1 - \frac{C_d}{C}\right)K^2}{4RC_d}. \tag{5.12}$$

From the study of the VDPO in Chap. 3, Sect. 3.2, we know that the amplitude general solution is given by:

$$I_{\mathrm{osc}}(t) = 2\sqrt{\frac{\delta_0}{\delta_2 + Be^{-2\delta_0 t}}}, \tag{5.13}$$

where $B = -\delta_2 + 4\delta_0/I_0^2$ is a constant that depends on the initial conditions and I_0 is the initial amplitude (i.e. for $t = 0$).

For steady state (i.e. $t \rightarrow \infty$), the oscillation amplitude obtained from Eq. (5.13) is

$$I_{\text{osc}} = 2\sqrt{\frac{\delta_0}{\delta_2}}. \tag{5.14}$$

Substituting the VDP parameters given by Eq. (5.12) into Eq. (5.14), we obtain the steady-state amplitude as a function of the circuit parameters:

$$I_{\text{osc}} \approx 2\sqrt{\frac{Rg_{m0}\left(1-\frac{C_d}{C}\right)-1}{3Rg_{m0}\left(1-\frac{C_d}{C}\right)K^2}} = \frac{2}{K}\sqrt{\frac{Rg_{m0}-\left(\frac{C}{C-C_d}\right)}{3Rg_{m0}}}, \tag{5.15}$$

The parameter K depends on the transistor working region. If the transistor is in strong inversion [5], K is given by:

$$K = \frac{1}{4I}. \tag{5.16}$$

If the transistor is in weak inversion [5], K is given by:

$$K \approx \frac{1}{I}. \tag{5.17}$$

If the transistor is working in a region between the strong and weak inversion, i.e. in moderate inversion, for which a model was proposed in [6], one can use the strong inversion model considering K as a fitting parameter.

To obtain the amplitude of the output voltage, which is easier to compare with the measurement results, we multiply Eq. (5.15) by the output resistance ($2R$) and obtain

$$V_{\text{osc}} \approx \frac{16R\left(1-\frac{C_d}{C}\right)I}{\sqrt{3}}\sqrt{\frac{Rg_{m0}-\left(\frac{C}{C-C_d}\right)}{Rg_{m0}}}. \tag{5.18}$$

The oscillation frequency is obtained from the third term on the left-hand side of Eq. (5.10) as:

$$\omega_0 \approx \sqrt{\frac{g_{m0}}{4RC_dC}} \tag{5.19}$$

The derived equations give approximate values for the amplitude and oscillation frequency in steady state. Below, we discuss the conditions necessary for the oscillation to start and the limit condition for avoiding clipping to maintaining the oscillator in the nearly sinusoidal regime.

5.2.1 Start-Up Conditions

An important aspect for the designer is the condition for the oscillation to start. The oscillation starts if the absolute value of the negative resistance exceeds the loss resistance R_{NL} that is generated by the oscillator's core transistors. This leads to the condition:

$$|R_{\mathrm{N}}| > R_{\mathrm{NL}}. \tag{5.20}$$

At the start-up ($i \approx 0$), the nonlinear resistance, R_{NL}, depends g_{m0}, the average transconductance of M_1 and M_2 . The negative resistance, R_{N}, depends on the oscillator capacitance C as can be seen in Eq. (5.7), and the condition ($C > C_d$) is necessary to have R_{N} negative. Thus, the conditions for the oscillation to start are [3]:

$$\begin{cases} R \cdot \left(1 - \dfrac{C_d}{C}\right) > \dfrac{1}{g_{m0}} & \text{(5.21a)} \\ C > C_d. & \text{(5.21b)} \end{cases}$$

From Eq. (5.21), one can see that the oscillator capacitance, C, must be higher than the parasitic capacitance C_d, and the negative resistance should be higher than the inverse of the average transconductance, g_{m0}. If the above conditions are met, the amplitude of oscillation grows exponentially and the nonlinear resistance also grows until in steady-state it matches the negative resistance. However, to ensure that the oscillator works in the near-sinusoidal regime an upper limit (for Rg_{m0}) must be set, otherwise the amplitude will not be limited by the circuit nonlinearities, but rather by the supply voltage V_{dd}. The upper limit is obtained by equating Eq. (5.15) to the maximum amplitude and solve with respect to Rg_{m0}. Thus, equating Eq. (5.15) to I, which is the maximum amplitude, using $K = 1/(4I)$ and solving with respect to Rg_{m0}, we obtain

$$R \cdot \left(1 - \frac{C_d}{C}\right) < \frac{1,049}{g_{m0}}. \tag{5.22}$$

The conditions in Eq. (5.21) define the minimum values for the oscillation to start and Eq. (5.22) gives the maximum value to avoid strong signal distortion (clipping).

5.2.2 Quality Factor

The Q of the circuit of Fig. 5.2 is calculated as:

$$Q = \frac{1}{R_{\rm NL}}\sqrt{\frac{L_e}{C}} = g_{m0}\sqrt{\frac{RC_d}{C}\frac{1}{g_{m0}}}. \tag{5.23}$$

Substituting Eq. (5.5) and Eq. (5.6) into Eq. (5.23) and considering that in steady-state $R\,(1 - C_d/C) \approx g_{m0}{}^{-1}$, we finally obtain

$$Q = \frac{Rg_{m0}}{C}\sqrt{C_d\,(C - C_d)}, \tag{5.24}$$

The maximum quality factor, $Q_{\rm max}$, is obtained by differentiating Eq. (5.24) with respect to C_d and equating the result to zero, that is:

$$\frac{\mathrm{d}Q}{\mathrm{d}C_d} = \frac{Rg_{m0}}{C}\frac{C - 2C_d}{2\sqrt{C_d\,(C - C_d)}} = 0. \tag{5.25}$$

From Eq. (5.25), one finds that:

$$Q_{\rm max} = \frac{Rg_{m0}}{2} \approx 1. \tag{5.26}$$

which is reached with $C = 2C_d$. This result is in accordance with that given in [7].

Thus, as a rule of thumb, the designer should enforce a floating capacitance, C, equal to twice the parasitics capacitance, C_d, (i.e. $C = 2C_d$) to guarantee the quality factor $Q \approx 1$. Hence, contrarily to the LC-oscillator, the Q of an RC-oscillator is limited to one and, therefore, Q cannot be used to reduce the phase noise.

The optimum values of circuit parameters guarantee the minimization of the phase noise [7]. With $Q \approx 1$, Eq. (5.21a) is reduced to $R > 2/g_{m0}$. Thus, for the oscillation to start, one has to ensure that R compensates the loss in the transconductance, g_{m0}. Moreover, to maximize the amplitude, the designer should use the maximum value for Rg_{m0} that is obtained from Eq. (5.22) (with $C = 2C_d$) resulting in:

$$Rg_{m0} \approx 2.1$$

Although, the above rule maximizes the amplitude for the nearly sinusoidal regime, for some applications the THD is still too high. In these cases, the designer should use the THD as a figure-of-merit to set the Rg_{m0} value. Moreover, a trade-off exists between the maximum amplitude and power consumption, since to maintain the desired frequency the ratio between R and g_{m0} must be constant. This trade-off can also be used to set the limit to Rg_{m0}.

Below, we present the design and simulation of an RC-oscillator (Fig. 5.1a), in which the design guidelines and a validation of the theoretical analysis by simulation are presented.

5.2.3 Design and Simulation Result

To confirm the theoretical analysis, we designed a 2.4-GHz RC-oscillator using a 130-nm standard CMOS technology considering ideal resistances, capacitances and current sources (circuit of Fig. 5.1a). To minimize the power requirements, we select the NMOS transistors with the largest W/L ratio available in the technology library ($W/L = 115.2\,\mu\text{m}/120\,\text{nm}$). From simulation, we obtain an approximated value for the parasitics capacitance of about $C_d = 172\,\text{fF}$. It follows that the floating capacitance must be $C = 2C_d = 344\,\text{fF}$. Substituting the frequency and capacitances values into Eq. (5.19) and using the conditions of Eq. (5.21), we obtain the minimum values for the resistances and transconductances as $R > 191.6\,\Omega$ and $g_{m0} > 10.4\,\text{mS}$. From the transconductance value, we obtain that the required bias current, I, based on the transistors dimension and technology parameters must be $I \approx 600\,\mu\text{A}$. With this low current, the transistors work in moderate inversion, and, hence, K is used as a fitting parameter.

The developed theoretical analysis neglects the gate currents. This can be done for low frequencies (or for high-bias currents). At high-frequency and with low-bias current, the gate current cannot be neglected and has a significant impact on the value of I_{osc}. For the designed oscillator, the gate current can be about 20% of the oscillator current, I_{osc}. Moreover, in the theory the transistors' output impedance and its impact on I_{osc} was neglected. However, as we show next, these approximations can be compensated by choosing the proper value of K.

The RC-oscillator with the above parameters was simulated to confirm the amplitude trend predicted by the theoretical analysis. The value of R was varied from 193 Ω to 210 Ω maintaining the other parameters constant. Simulation results show the amplitude (Fig. 5.3) and frequency change (Fig. 5.4) with respect to R. For the amplitude, a comparison between the simulation results (black dots) and the theoretical analysis is shown in Fig. 5.3. The dashed curve represents the values predicted by Eq. (5.18), using the strong inversion K, and the solid curve the values predicted by Eq. (5.18) with $K = 1.41/(4I)$. It can be seen from Fig. 5.3 that the oscillation amplitude increases with R as predicted by the theoretical analysis. The approximation using the strong inversion K is poor, because the transistors are in moderate inversion. Yet, the increase of K by 41% gives a better agreement between simulation and theory (Fig. 5.3, solid curve).

Figure 5.4 shows the dependence of frequency on R; the simulation results are represented by the black dots, and the solid curve represents the values predicted by Eq. (5.19). Although, the frequency trend is in agreement with the theoretical analysis, the values have a significant difference. Investigating further, we concluded that neglecting the channel-length modulation explains this difference.

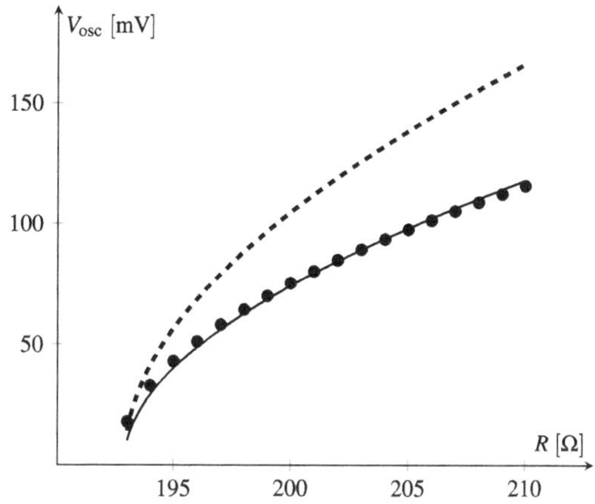

Fig. 5.3 Oscillation amplitude

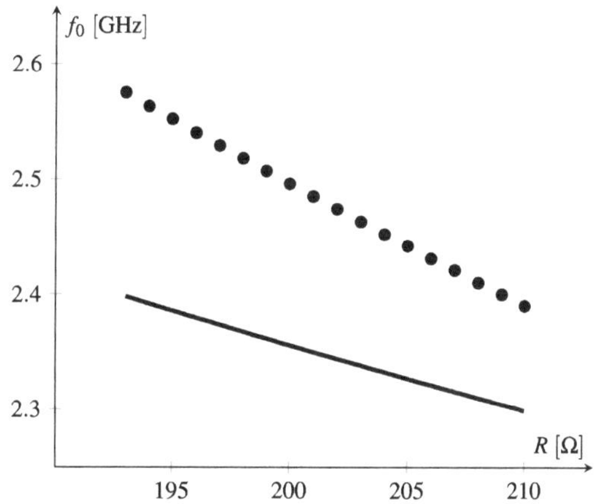

Fig. 5.4 Oscillation frequency

The simulation results show that the RC-oscillator in the nearly sinusoidal regime can be approximated by the series VDPO (Fig. 5.2). Moreover, if we consider the nonlinearities of the core transistors, a large-signal analysis can be avoided.

5.3 Quadrature RC-Oscillator

In this section, we analyse an oscillator with quadrature outputs using two RC-oscillators. Two transconductance amplifiers are used to couple the oscillators, as shown in Fig. 5.5. The coupling forces the oscillators to synchronize and oscillate at the same frequency, as shown in Chap. 4, where locking of an oscillator to an external signal is studied. The synchronization process of coupled oscillators is equivalent to the synchronization of injection-locked oscillators: in coupled oscillators, the locking signal for one oscillator is the output of the other oscillator.

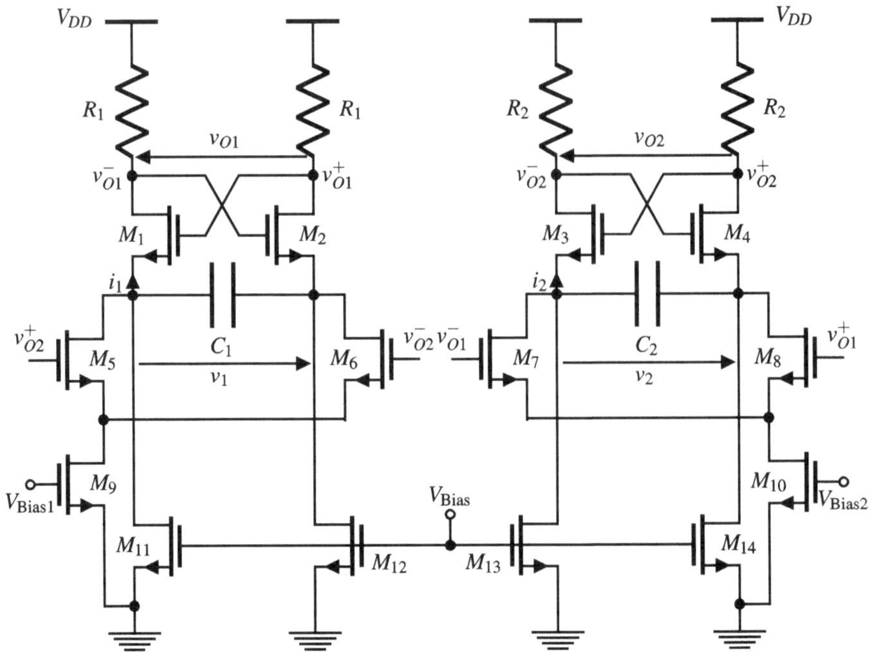

Fig. 5.5 Quadrature oscillator with active coupling

Thus, coupling ensures frequency synchronization, but the oscillators can lock either in-phase or in quadrature. To obtain quadrature outputs, it is necessary to cross-couple the oscillators (Fig. 5.5); direct coupling synchronizes both oscillators but generates in-phase outputs. Here, we focus on cross-coupling, since we are interested in quadrature signal generators.

The circuit implementation of a quadrature oscillator is shown in Fig. 5.5. Resistances R_1 and R_2, and the current mirrors implemented by transistors M_{11} to M_{14} set the oscillator bias point. The differential transconductance amplifiers are implemented by the differential pairs composed of transistors M_5, M_6 and M_7, M_8, and their tail current sources M_9 and M_{10}, respectively.

The circuit analysis is done as follows. First, the small-signal equivalent circuit is obtained. From it, the differential equations that govern the system are derived. These differential equations are used to obtain both the transient and steady-state performance of the circuit. Then, assuming that there are no mismatches between the oscillators, the key parameters, frequency, amplitude and phase of the quadrature oscillator are derived for the steady state. After that, using the steady-state results (equilibrium points), a stability analysis of each equilibrium point is presented. In the last subsection, considering mismatches between the oscillators, the amplitude- and phase-error equations are derived.

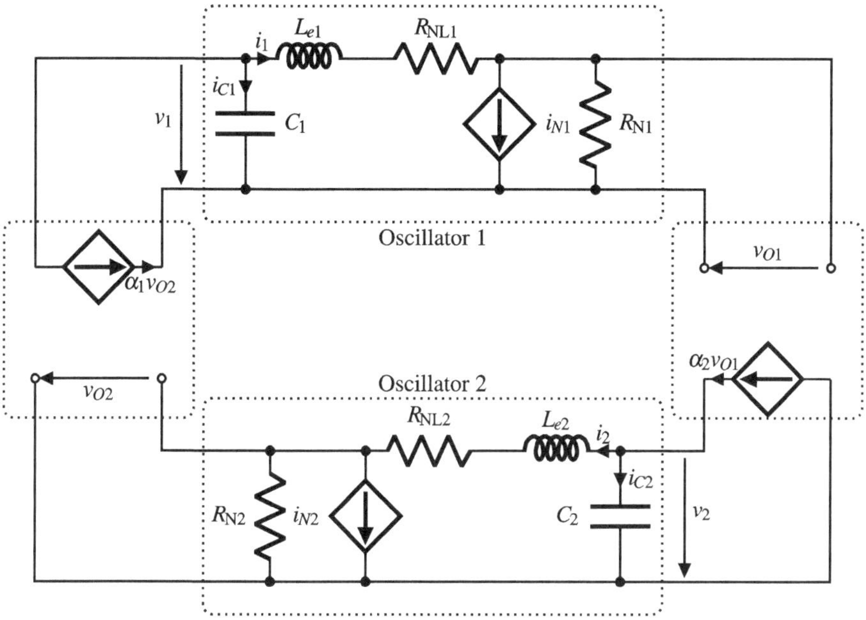

Fig. 5.6 Active coupling small-signal equivalent circuit

5.3.1 *Incremental Model*

The small-signal equivalent circuit is obtained by substituting each RC-oscillator by its equivalent circuit (Fig. 5.2) and the transconductance amplifiers by current sources controlled by the output voltage of each oscillator, as shown in Fig. 5.6.

The circuit of Fig. 5.6 consists of two series VDPOs, each driven by a current source. Using the driven series VDPO differential equation (4.22) (derived in Sect. 4.3.1) applied to each oscillator, we obtain

$$\begin{cases} \dfrac{d^2 i_1}{dt^2} - 2\left(\delta_0 - \delta_2 i_1^2\right)\dfrac{di_1}{dt} + \omega_1^2 i = 2R_2\left(1 - \dfrac{C_{d2}}{C_2}\right)\omega_1^2 \alpha_2 i_2 & (5.27a) \\ \dfrac{d^2 i_2}{dt^2} - 2\left(\gamma_0 - \gamma_2 i_2^2\right)\dfrac{di_2}{dt} + \omega_2^2 i = -2R_1\left(1 - \dfrac{C_{d1}}{C_1}\right)\omega_2^2 \alpha_1 i_1 & (5.27b) \end{cases}$$

where ω_i ($i = 1, 2$) is the free-running frequency, and α_i is the coupling factor of the ith oscillator. The parameters of the first VDPO are given by:

$$\delta_0 = \frac{R_1 g_{m1}\left(1 - \frac{C_{d1}}{C_1}\right) - 1}{4R_1 C_{d1}}, \quad \delta_2 = \frac{3R_1 g_{m1}\left(1 - \frac{C_{d1}}{C_1}\right) K^2}{4R_1 C_{d1}}. \tag{5.28}$$

and the parameters of the second oscillator are given by:

$$\gamma_0 = \frac{R_2 g_{m2}\left(1 - \frac{C_{d2}}{C_2}\right) - 1}{4R_2 C_{d2}}, \quad \gamma_2 = \frac{3R_1 g_{m1}\left(1 - \frac{C_{d1}}{C_1}\right) K^2}{4R_1 C_{d1}}. \tag{5.29}$$

To solve the differential equations (5.27), we use the harmonic balance method [8], with the assumptions of slowly varying amplitude and phase, and neglecting the high-order terms. Thus, the solutions have the form:

$$\begin{cases} i_1(t) = I_{o1}(t)\sin(\omega t - \phi_1) & (5.30a) \\ i_2(t) = I_{o2}(t)\sin(\omega t - \phi_2). & (5.30b) \end{cases}$$

where I_{oi} is the current amplitude, ϕ_i is the phase of the ith oscillator and ω is the common angular frequency of oscillation. Note that we are assuming that both oscillators are working at the same frequency but with different phases. Using the harmonic balance results in the system of four first-order differential equations:

$$\begin{cases} \frac{\mathrm{d}I_1}{\mathrm{d}t} = \delta_0 I_1 - \frac{1}{4}\delta_2 I_1^3 - \frac{\alpha_{K1}}{2\omega} I_2 \sin\Delta\phi & (5.31a) \\ \frac{\mathrm{d}I_2}{\mathrm{d}t} = \gamma_0 I_2 - \frac{1}{4}\gamma_2 I_2^3 - \frac{\alpha_{K2}}{2\omega} I_1 \sin\Delta\phi & (5.31b) \\ \frac{\mathrm{d}\phi_1}{\mathrm{d}t} = \frac{\omega^2 - \omega_1^2}{2\omega} + \frac{\alpha_{K1}}{2\omega}\frac{I_2}{I_1}\cos\Delta\phi & (5.31c) \\ \frac{\mathrm{d}\phi_2}{\mathrm{d}t} = \frac{\omega^2 - \omega_2^2}{2\omega} - \frac{\alpha_{K2}}{2\omega}\frac{I_1}{I_2}\cos\Delta\phi & (5.31d) \end{cases}$$

where $\Delta\phi = \phi_2 - \phi_1$ is the phase difference and ω_i is the free-running frequency of the ith oscillator. The coupling parameters, α_{Ki}, are given by:

$$\alpha_{K1} = 2R_2\left(1 - \frac{C_{d2}}{C_2}\right)\omega_1^2\alpha_1, \qquad \alpha_{K2} = 2R_1\left(1 - \frac{C_{d1}}{C_1}\right)\omega_2^2\alpha_2. \tag{5.32}$$

Analysing Eq. (5.31), one can obtain both the steady state and the transient performance.

5.3.2 *Quadrature Oscillator Without Mismatches*

To determine the phase, amplitude and frequency of oscillators, we consider that there are no mismatches between their components, i.e. $R_1 = R_2 = R$, $C_1 = C_2 = C$, $L_{e1} = L_{e1} = L_e$ and $\alpha_1 = \alpha_2 = \alpha$. The free-running frequencies are also equal, i.e. $\omega_1 = \omega_2 = \omega_0$, and the VDP parameters are equal as well; $\delta_0 = \gamma_0$ and

$\delta_2 = \gamma_2$, and the coupling factors $\alpha_{K1} = \alpha_{K2} = \alpha_K$. At the steady state, for which $\mathrm{d}I_1/\mathrm{d}t = \mathrm{d}I_2/\mathrm{d}t = \mathrm{d}\phi_1/\mathrm{d}t = \mathrm{d}\phi_2/\mathrm{d}t = 0$, Eq. (5.31) is reduced to:

$$\begin{cases} \delta_0 I_1 - \dfrac{1}{4}\delta_2 I_1^3 - \dfrac{\alpha_K}{2\omega} I_2 \sin\Delta\phi = 0 & (5.33\text{a}) \\ \delta_0 I_2 - \dfrac{1}{4}\delta_2 I_2^3 - \dfrac{\alpha_K}{2\omega} I_1 \sin\Delta\phi = 0 & (5.33\text{b}) \\ \dfrac{\omega^2 - \omega_0^2}{2\omega} + \dfrac{\alpha_K}{2\omega}\dfrac{I_1}{I_2}\cos\Delta\phi = 0 & (5.33\text{c}) \\ \dfrac{\omega^2 - \omega_0^2}{2\omega} - \dfrac{\alpha_K}{2\omega}\dfrac{I_2}{I_1}\cos\Delta\phi = 0 & (5.33\text{d}) \end{cases}$$

Three solutions are possible. A zero amplitude for both oscillators, i.e. $I_{o1} = I_{o2} = 0$, satisfies the equations. Note that to avoid the indeterminate multiplier of $0/0$, one can multiply I_1 by Eq. (5.33c) and I_2 by Eq. (5.33d). The second and third solutions with equal amplitudes $I_1 = I_2 = I_{osc}$ and with quadrature outputs $\Delta\phi = \pi/2$ and $\Delta\phi = -\pi/2$, respectively, satisfy the equations. The second and third solutions are obtained by dividing Eq. (5.33a) by I_1, and Eq. (5.33b) by I_2 and subtracting Eq. (5.33a) from Eq. (5.33b). This results in:

$$-\frac{1}{4}\delta_2\left(I_2^2 - I_1^2\right) - \frac{\alpha_K}{2\omega}\left(\frac{I_1}{I_2} - \frac{I_2}{I_1}\right)\sin\Delta\phi = 0. \tag{5.34}$$

One can see from Eq. (5.34) that the case of equal amplitudes, i.e. $|I_1| = |I_2|$, satisfies the equation. Then, for equal amplitudes, subtraction of Eq. (5.33c) from Eq. (5.33d) results in:

$$-\alpha_K \cos\Delta\phi = 0. \tag{5.35}$$

Thus, only the phase differences $\Delta\phi = \pi/2$ and $\Delta\phi = -\pi/2$ satisfy Eq. (5.35).

Solving Eq. (5.33a), or Eq. (5.33b), with respect to I_{osc}, considering equal amplitudes $I_1 = I_2 = I_{osc}$, we obtain

$$I_{osc} = 2\sqrt{\frac{\delta_0 \pm \frac{\alpha_K}{2\omega}}{\delta_2}}, \tag{5.36}$$

Then, subtracting Eq. (5.33c) from Eq. (5.33d) and solving with respect to $\Delta\phi$, we obtain

$$\Delta\phi = \pm\frac{\pi}{2}. \tag{5.37}$$

Moreover, combining Eq. (5.33c) and Eq. (5.33d) results in:

$$\omega = \omega_0. \tag{5.38}$$

These results lead us to the conclusion that, with no mismatches between the oscillators, at the steady state, the outputs are in perfect quadrature. The oscillation frequency is equal to the free-running frequency. The oscillation amplitude is different from that of a single RC-oscillator. Moreover, the amplitude has two modes, described by Eq. (5.36). The mechanism by which a particular mode is selected is discussed in the next section.

5.3.3 Stability of the Equilibrium Points

To analyse the stability of the steady-state oscillations, we determine the stability of each equilibrium point. This is done by analysing the characteristic equation and its eigenvalues. Furthermore, to understand how the circuit reaches the steady state, the paths in the vicinity of each equilibrium point are drawn.

Simplifying Eq. (5.31) by combining Eq. (5.31c) and Eq. (5.31d), we obtain

$$\begin{cases} \dfrac{\mathrm{d}I_1}{\mathrm{d}t} = \delta_0 I_1 - \dfrac{1}{4}\delta_2 I_1^3 - \dfrac{\alpha_K}{2\omega} I_2 \sin \Delta\phi & \text{(5.39a)} \\ \dfrac{\mathrm{d}I_2}{\mathrm{d}t} = \delta_0 I_2 - \dfrac{1}{4}\delta_2 I_2^3 - \dfrac{\alpha_K}{2\omega} I_1 \sin \Delta\phi & \text{(5.39b)} \\ \dfrac{\mathrm{d}\Delta\phi}{\mathrm{d}t} = -\dfrac{\alpha_K}{2\omega}\left(\dfrac{I_1}{I_2} - \dfrac{I_2}{I_1}\right) \cos \Delta\phi & \text{(5.39c)} \end{cases}$$

From Eq. (5.39c), one finds that $\mathrm{d}\Delta\phi/\mathrm{d}t = 0$ when the quadrature between outputs is reached. Thus, if we assume that the outputs are in quadrature $\Delta\phi = \pi/2$, and Eq. (5.39) can be reduced to:

$$\begin{cases} \dfrac{\mathrm{d}I_1}{\mathrm{d}t} = \delta_0 I_1 - \dfrac{1}{4}\delta_2 I_1^3 - \dfrac{\alpha_K}{2\omega} I_2 & \text{(5.40a)} \\ \dfrac{\mathrm{d}I_2}{\mathrm{d}t} = \delta_0 I_2 - \dfrac{1}{4}\delta_2 I_2^3 - \dfrac{\alpha_K}{2\omega} I_1 & \text{(5.40b)} \end{cases}$$

Note that the differential equations are nonlinear, which leads to a cumbersome analysis. For convenience, we analyse the transient in the vicinity of the equilibrium point, where the system can be linearized. To linearize Eq. (5.40), for any point in the phase plane, we calculate its Jacobian matrix:

$$J = \begin{bmatrix} \delta_0 - \frac{3}{4}\delta_2 I_{\mathrm{osc}}^2 & -\frac{\alpha_K}{2\omega} \\ -\frac{\alpha_K}{2\omega} & \delta_0 - \frac{3}{4}\delta_2 I_{\mathrm{osc}}^2 \end{bmatrix}, \tag{5.41}$$

where I_{osc} is the steady-state amplitude at the equilibrium points. To define the stability of an equilibrium point and determine the phase portrait near it, we must determine the characteristic equation:

$$\lambda^2 - T\lambda + D = 0, \tag{5.42}$$

where T is the trace (sum of the main diagonal elements) and D is the determinant of the Jacobian matrix. The conditions for stability are $T < 0$ and $D > 0$ [8]. For $T > 0$ or $D < 0$, the equilibrium point is unstable [8].

The steady-state analysis produces three equilibrium points: one is at the origin $E_0 = (0, 0)$, the other two are $E_1 = (I_{osc}, I_{osc})$ and $E_2 = (-I_{osc}, -I_{osc})$, in the first and third quadrants, respectively. At the equilibrium point E_0, the amplitudes are equal ($I_1 = I_2 = 0$) and, therefore, the Jacobian matrix is

$$J_{E_0} = \begin{bmatrix} \delta_0 & -\frac{\alpha_K}{2\omega} \\ -\frac{\alpha_K}{2\omega} & \delta_0 \end{bmatrix}, \tag{5.43}$$

From Eq. (5.43), we obtain the characteristic equation:

$$\lambda^2 - 2\delta_0\lambda + \left(\delta_0^2 - \frac{\alpha_K^2}{4\omega^2}\right) = 0,$$

where the trace $T = 2\delta_0 > 0$ since δ_0 is positive, which means that the equilibrium point E_0 is unstable. If we consider that $|\delta_0| < |\alpha_K|$, the determinant of Eq. (5.43) is negative meaning that we have a saddle point at E_0. If $|\delta_0| > |\alpha_K|$, we have an unstable node.

At the equilibrium point E_1, the amplitudes are equal $I_1 = I_2 = I_{osc}$. Substituting Eq. (5.36) into Eq. (5.41), the Jacobian for E_1 is obtained:

$$J_{E_1} = \begin{bmatrix} -2\delta_0 & -\frac{\alpha_K}{2\omega} \\ -\frac{\alpha_K}{2\omega} & -2\delta_0 \end{bmatrix}. \tag{5.44}$$

The characteristic equation of the matrix in Eq. (5.44) is

$$\lambda^2 + 4\delta_0\lambda + \left(4\delta_0^2 - \frac{\alpha_K^2}{4\omega^2}\right) = 0, \tag{5.45}$$

where $T < 0$ and $D > 0$ since δ_0 is positive. This means that the equilibrium point E_1 is stable. The eigenvalues are

$$\lambda = -\frac{T}{2} \pm \frac{1}{2}\sqrt{T^2 - 4D} = -\frac{T}{2} \pm \frac{1}{2}\sqrt{\Delta}.$$

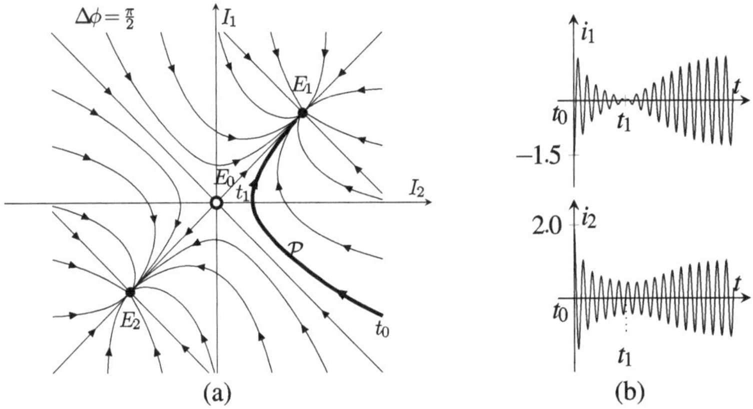

Fig. 5.7 Series active coupling (**a**) phase portrait and (**b**) time solution

If $\Delta > 0$, the eigenvalues are real. For E_1, substituting T and D of J_{E_1} into Δ, results in:

$$\Delta = \frac{\alpha_K^2}{\omega^2}.$$

Since $\alpha_K^2/\omega^2 > 0$, we can conclude that at E_1 we have a stable node. The same conclusion can be drawn for E_2 since its Jacobian matrix is equal to J_{E_1}, as shown by the phase portrait in Fig. 5.7a. The stable and unstable equilibrium points are represented, in Fig. 5.7a, as black and white circles, respectively.

The phase portrait of Fig. 5.7a is a graphic representation of the amplitudes' evolution. Consider, for instance, the path $\mathcal{P}$ (bold line) that corresponds to the initial conditions: $i_1(t_0) = -1.5\,\text{mA}$, $i_2(t_0) = 2\,\text{mA}$ and $\Delta\phi = -\pi/2$. The phase portrait shows that, from this initial point, the amplitude I_2 decreases until it reaches its minimum value, at $t = t_1$. For $t > t_1$, the amplitude I_2 increases until it reaches the equilibrium point. The amplitude I_1 increases until it reaches the steady state at E_1 passing from negative to positive values at $t = t_1$. This behaviour can be seen in the time-domain representations (Fig. 5.7b).

5.3.4 *Quadrature Oscillator with Mismatches*

In this subsection, we analyse the impact of the mismatches in the passive components on the amplitude and phase of single RC-oscillators. We assume that the oscillators' core transistors are identical, and, therefore, their transconductances and capacitances are equal.

In the following derivation, we also consider that the oscillators have reached the steady state and that the mismatches are small, i.e. $\Delta R/R < 1\%$ and $\Delta C/C < 1\%$.

To simplify the derivation, we neglect the terms with a multiplication of relative mismatches, e.g. we consider that $\Delta R \Delta C/(RC) \approx 0$. The error equations are derived as functions of the mismatches of R and C.

For $\omega_1 \neq \omega_2$, $\gamma_0 \neq \delta_0$, $\gamma_2 \neq \delta_2$ and $\alpha_{K1} \neq \alpha_{K2}$ at the steady state, Eq. (5.31) is reduced to:

$$\begin{cases} \delta_0 - \frac{1}{4}\delta_2 I_1^2 - \frac{\alpha_{K1}}{2\omega}\frac{I_2}{I_1}\sin\Delta\phi = 0 & \text{(5.46a)} \\ \gamma_0 - \frac{1}{4}\gamma_2 I_2^2 - \frac{\alpha_{K2}}{2\omega}\frac{I_1}{I_2}\sin\Delta\phi = 0 & \text{(5.46b)} \\ \frac{\omega^2 - \omega_1^2}{2\omega} + \frac{\alpha_{K1}}{2\omega}\frac{I_2}{I_1}\cos\Delta\phi = 0, & \text{(5.46c)} \\ \frac{\omega^2 - \omega_2^2}{2\omega} - \frac{\alpha_{K2}}{2\omega}\frac{I_1}{I_2}\cos\Delta\phi = 0, & \text{(5.46d)} \end{cases}$$

By adding Eq. (5.46c) and Eq. (5.46d), one obtains the equations for the oscillation frequency:

$$2\omega^2 = \left(\omega_1^2 + \omega_2^2\right) - \left(\alpha_{K1}\frac{I_2}{I_1} - \alpha_{K2}\frac{I_1}{I_2}\right)\cos\Delta\phi, \tag{5.47}$$

For small mismatches and small coupling factor, then Eq. (5.47) is reduced to:

$$\omega \approx \sqrt{\frac{\omega_1^2 + \omega_2^2}{2}}. \tag{5.48}$$

This equation shows that the oscillation frequency of the quadrature oscillator is the quadratic mean of the free-running frequencies of the *RC*-oscillators.

To obtain the phase-error equation, one should subtract Eq. (5.46c) from Eq. (5.46d). This results in:

$$\left(\alpha_{K2}\frac{I_1}{I_2} + \alpha_{K1}\frac{I_2}{I_1}\right)\cos\Delta\phi = \omega_1^2 - \omega_2^2, \tag{5.49}$$

The phase difference $\Delta\phi$ can be written as $\Delta\phi = \pi/2 + \varepsilon_\phi$, where ε_ϕ is the phase error. Using the trigonometrical relation $\cos(\Delta\phi) = -\sin(\varepsilon_\phi)$, assuming small phase error $-\sin(\varepsilon_\phi) \approx -\varepsilon_\phi$ in Eq. (5.49) and solving it with respect to the phase error result in:

$$\varepsilon_\phi \approx -\frac{\omega_1^2 - \omega_2^2}{\alpha_{K2}\frac{I_1}{I_2} + \alpha_{K1}\frac{I_2}{I_1}}, \tag{5.50}$$

The mismatched amplitudes may be described as:

$$I_1 = I_{\text{osc}}\left(1 - \frac{\varepsilon_A}{2}\right), \qquad I_2 = I_{\text{osc}}\left(1 + \frac{\varepsilon_A}{2}\right). \tag{5.51}$$

where ε_A is the amplitude error. Substituting Eq. (5.51) in Eq. (5.50), we obtain

$$\varepsilon_\phi \approx -\frac{\omega_1^2 - \omega_2^2}{(\alpha_{K1} + \alpha_{K2}) + (\alpha_{K1} - \alpha_{K2})\,\varepsilon_A}, \tag{5.52}$$

Considering that $C_{d1} = C_{d2} = C_d$, $g_{m1} = g_{m2} = g_m$ and using $R_1 = R\,(1 - \Delta R/(2R))$, $R_2 = R\,(1 + \Delta R/(2R))$, $C_1 = C\,(1 - \Delta C/(2C))$ and $C_2 = C\,(1 - \Delta C/(2C))$, one finds that the difference between the free-running frequencies of each RC-oscillator is given by:

$$\omega_1^2 - \omega_2^2 = \omega_0^2\left(\frac{\Delta R}{R} + \frac{\Delta C}{C}\right). \tag{5.53}$$

Using the same assumptions, the term $\alpha_{K1} - \alpha_{K2}$ is calculated as:

$$\alpha_{K1} - \alpha_{K2} \approx -2R\omega_0^2\left[\left(1 - \frac{C_d}{C}\right)\frac{\Delta\alpha}{\alpha} + \left(\alpha\frac{C_d}{C}\right)\frac{\Delta R}{R}\right] \tag{5.54}$$

and the term $\alpha_{K1} + \alpha_{K2}$ is given by:

$$\alpha_{K1} + \alpha_{K2} \approx 4R\omega_0^2\alpha\left(1 - \frac{C_d}{C}\right) \tag{5.55}$$

Substituting Eq. (5.53), Eq. (5.54) and Eq. (5.55) into Eq. (5.52) results in:

$$\varepsilon_\phi \approx -\frac{\frac{\Delta R}{R} + \frac{\Delta C}{C}}{4R\left(1 - \frac{C_d}{C}\right)\alpha - 2R\left(1 - \frac{C_d}{C}\right)\frac{\Delta\alpha}{\alpha}\varepsilon_A - 2R\frac{C_d}{C}\alpha\frac{\Delta R}{R}\varepsilon_A} \tag{5.56}$$

Then, for small mismatches $\Delta\alpha/\alpha \ll \alpha$ and $\Delta R/R\varepsilon_A \ll 1$, Eq. (5.56) is reduced to:

$$\varepsilon_\phi \approx -\frac{1}{4R\left(1 - \frac{C_d}{C}\right)\alpha}\left(\frac{\Delta R}{R} + \frac{\Delta C}{C}\right) \tag{5.57}$$

Equation (5.57) gives the phase error in radians. To obtain the phase error in degrees, one has to multiply Eq. (5.57) by $180/\pi$ to obtain

$$\varepsilon_{\phi(\text{deg})} \approx -\frac{180}{\pi}\frac{1}{4R\left(1 - \frac{C_d}{C}\right)\alpha}\left(\frac{\Delta R}{R} + \frac{\Delta C}{C}\right) \tag{5.58}$$

Hence, one can see that the phase error is directly proportional to the mismatches and inversely proportional to the coupling factor.

5.4 Simulation Results

The circuit of Fig. 5.5 was simulated using standard 130 nm CMOS technology parameters. The circuit parameters are: $C_1 = C_2 = C = 77$ fF, $R_1 = R_2 = R = 600\,\Omega$, $(W/L) = 115.2\,\mu$m/120 nm for transistors M_1,M_2,M_3,M_4,M_9 and M_{10}, $(W/L) = 14.4\,\mu$m/120 nm for M_5,M_6,M_7 and M_8, $I = 0.6$ mA, $I_{cp} = 100\,\mu$A, and the supply voltage is 1.2 V. The voltage and current sources are assumed to be ideal.

Simulating the circuit with component mismatches from -2% to $+2\%$, we obtain the amplitude and phase errors shown in Fig. 5.8 and Fig. 5.9, respectively.

In Fig. 5.9, the solid line is the plot of the theoretical phase error given by Eq. (5.58). This line represents the phase error with respect to one of the mismatches considering the other equal to zero.

The simulation results show that the phase error with respect to the capacitance mismatches agrees well with the theory, as shown in Fig. 5.9. However, for the resistance mismatch, the simulation results diverge slightly (higher slope) from the

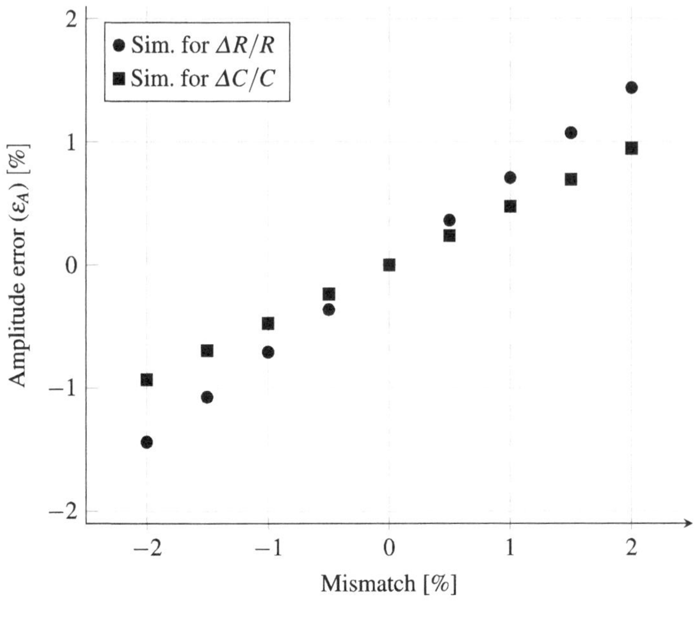

Fig. 5.8 Impact of the resistance and capacitance mismatches on the amplitude error, using the circuit parameters: $R = 210\,\Omega$, $I = 600\,\mu$A and $I_{cp} = 100\,\mu$A ($\alpha = g_{m0} \approx 0.758$ mS)

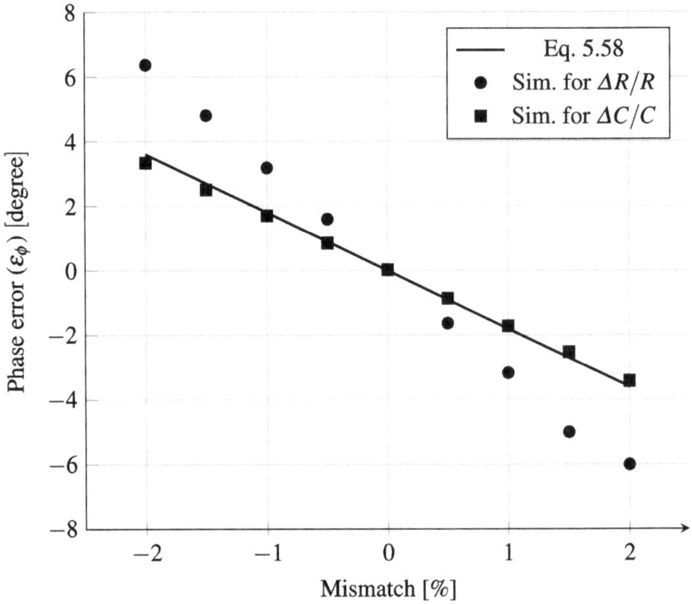

Fig. 5.9 Impact of the resistance and capacitance mismatches on the phase error, using the circuit parameters: $R = 210\,\Omega$, $I = 600\,\mu A$ and $I_{cp} = 100\,\mu A$ ($\alpha = g_{m0} \approx 0.758\,mS$)

theory. This deviation is explained by the fact that the drain-to-source dynamic resistance of the transistors was neglected.

To determine the impact of the coupling factor on the phase error, we simulate the circuit with a constant mismatch of 2% and sweep the coupling factor. The results (Fig. 5.10) show that the phase error is inversely proportional to the coupling factor. These results show a significant deviation from the theory. This could be due to neglecting the drain-to-source dynamic resistance and different values of parasitic capacitances. Due to the Miller effect, the input capacitances of the transconductance amplifiers, used in coupling, increase with the increase of the coupling factor. Thus, the input capacitances of the transconductance amplifiers load the circuit, meaning that the capacitance C_d increases and opposes the phase-error reduction. Another consequence of the C_d increase is the decrease of the oscillation frequency.

5.5 Conclusions

In this chapter, we presented the study of the active coupling quadrature RC-oscillator. It was shown that for the sinusoidal regime, this quadrature oscillator can be modeled as two VDPOs coupled by two transconductors. First, it was shown that

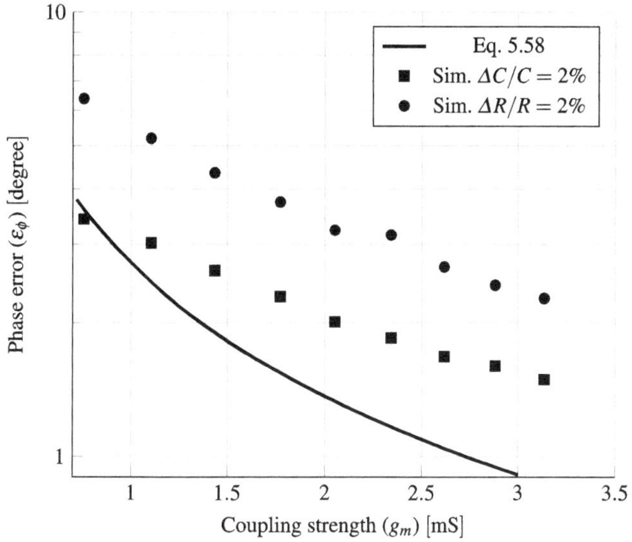

Fig. 5.10 Phase error as a function of the coupling factor, using the circuit parameters: $R = 210\,\Omega$ and $I = 600\,\mu\text{A}$

the single RC-oscillator can be modeled by the VDPO. The relationships between the circuit parameters and the VDPO parameters were derived and confirmed by simulation. Next, the incremental circuit of the quadrature oscillator was obtained by substituting each single RC-oscillator by a VDPO. Then, the transient and steady-state performance of the quadrature oscillator was studied and the equations of the oscillator key parameters, oscillation frequency and phase error were derived and validated by simulation.

We found that the oscillation frequency is insensitive to the mismatches and is given by the quadratic mean between the free-running frequencies of the coupled oscillators. However, contrary to what the theory predicted, simulations reveal that the oscillation frequency depends on the coupling factor. The increase of the coupling factor decreases the oscillation frequency. This is explained by the fact that the input capacitances of the transconductance amplifiers depend on the coupling factor. Due to the Miller effect, the input capacitances of the transconductance amplifiers depend on the bias current. Since the input capacitances are in parallel with the oscillator capacitance, they influence the oscillation frequency.

References

1. L. Oliveira, J. Fernandes, I. Filanovsky, C. Verhoeven, M. Silva, *Analysis and Design of Quadrature Oscillators* (Springer, Heidelberg, 2008)
2. L. Romano, S. Levantino, C. Samori, A.L. Lacaita, Multiphase LC oscillators. IEEE Trans. Circuits Syst. I **53**(7), 1579–1588 (2006)

3. A. Buonomo, A. Lo Schiavo, Analysis of emitter (source)-coupled multivibrators. IEEE Trans. Circuits Syst. I **53**(6), 1193–1202 (2006)
4. I.M. Filanovsky, C.J. Verhoeven, Sinusoidal and relaxation oscillations in Source-Coupled multivibrators. IEEE Trans. Circuits Syst. II **54**(11), 1009–1013 (2007)
5. D.A. Johns, K. Martin, *Analog Integrated Circuit Design*, 2nd edn. (Wiley, New York, 2011)
6. I.M. Filanovsky, J.K. Järvenhaara, N.T. Tchamov, On moderate inversion/saturation regions as approximations to "reconciliation" model, in *2016 IEEE Canadian Conference on Electrical and Computer Engineering (CCECE)* (2016), pp. 1–5
7. B. Razavi, *RF Microelectronics* (Prentice-Hall, Upper Saddle River, 1998)
8. D.W. Jordan, P. Smith, *Nonlinear Ordinary Differential Equations: An Introduction for Scientists and Engineers*, 4th edn. (Oxford University Press, New York, 2007)

Chapter 6
Capacitive Coupling RC-Oscillator

6.1 Introduction

In the previous chapter, the quadrature oscillator with active coupling was analysed. The disadvantages of this coupling are the increase of the noise and power dissipation. In this chapter, we analyse the passive coupling which is an alternative method that diminishes these disadvantages. In passive coupling, the amplifiers are substituted by passive elements (usually inductors or capacitors). The coupling based on inductors [1] and transformers [2, 3] has a larger area than the active coupling. Using capacitive coupling in LC-oscillators does not result in minimal area and, in addition, reduces the oscillation frequency [4, 5].

Here, the quadrature RC-oscillator with capacitive coupling is investigated [6]. The capacitive coupling is noiseless and requires a small area. Since the coupling capacitors do not add noise, we expect 3-dB phase-noise improvement (due to the coupling), with a marginal increase of the dissipation power, and a figure-of-merit (FoM) comparable to that of the best state-of-the-art RC-oscillators. Contrarily to what may be expected, with the increase of the coupling capacitances (stronger coupling) the oscillation frequency increases [6]. We present a theory that explains this behaviour and we derive the equations of the frequency, phase error, and amplitude mismatch. These equations are validated by simulation. The theory shows that both phase and amplitude errors are reduced with the increase of the coupling factor. Moreover, the phase error is proportional to the amplitude mismatch, indicating that an automatic phase-error minimization based on the amplitude mismatch reduction is possible. The theory shows also that the phase noise has a low sensitivity to the coupling factor. We also study bimodal oscillations and phase ambiguity for this type of coupling and compare it with other works [7]. To validate the theory, a 2.4-GHz quadrature voltage-controlled oscillator (QVCO) based on two RC-oscillators with capacitive coupling was fabricated, in UMC 130 nm CMOS process.

J. C. F. d. A. Casaleiro et al., *Quadrature RC–Oscillators*, Analog Circuits and Signal Processing, https://doi.org/10.1007/978-3-030-00740-9_6

The chapter is organized as follows. We first present the circuit implementation and its incremental model using the VDP approximation. After that, we present the analysis of the oscillator with capacitive coupling and derive the equations for the oscillation frequency, phase, and amplitude. The stability analysis is included, extending the analysis presented in [8]. Equations for the phase error and amplitude mismatch are derived, relating these with the circuit parameters, extending the results presented in [6]. Following the theoretical analysis, simulation results are presented and the comparison with theory is done. At the end of the chapter, the experimental results followed by the conclusions are presented. The experimental results are compared with the state-of-the-art of nearly sinusoidal RC-oscillators. The conclusions highlight the inverse proportionality between the errors and the coupling factor and the low sensitivity of the phase noise with respect to the coupling factor.

6.2 Quadrature Oscillator

In a quadrature RC-oscillator with capacitive coupling, two RC-oscillators are coupled by four coupling capacitances, C_X, as shown in Fig. 6.1. It is worth mentioning that quadrature outputs are obtained with cross-coupling (Fig. 6.1). Transistors M_{1-4} are the oscillator core. M_9, M_{11}, M_{12} and M_{14} are current sources

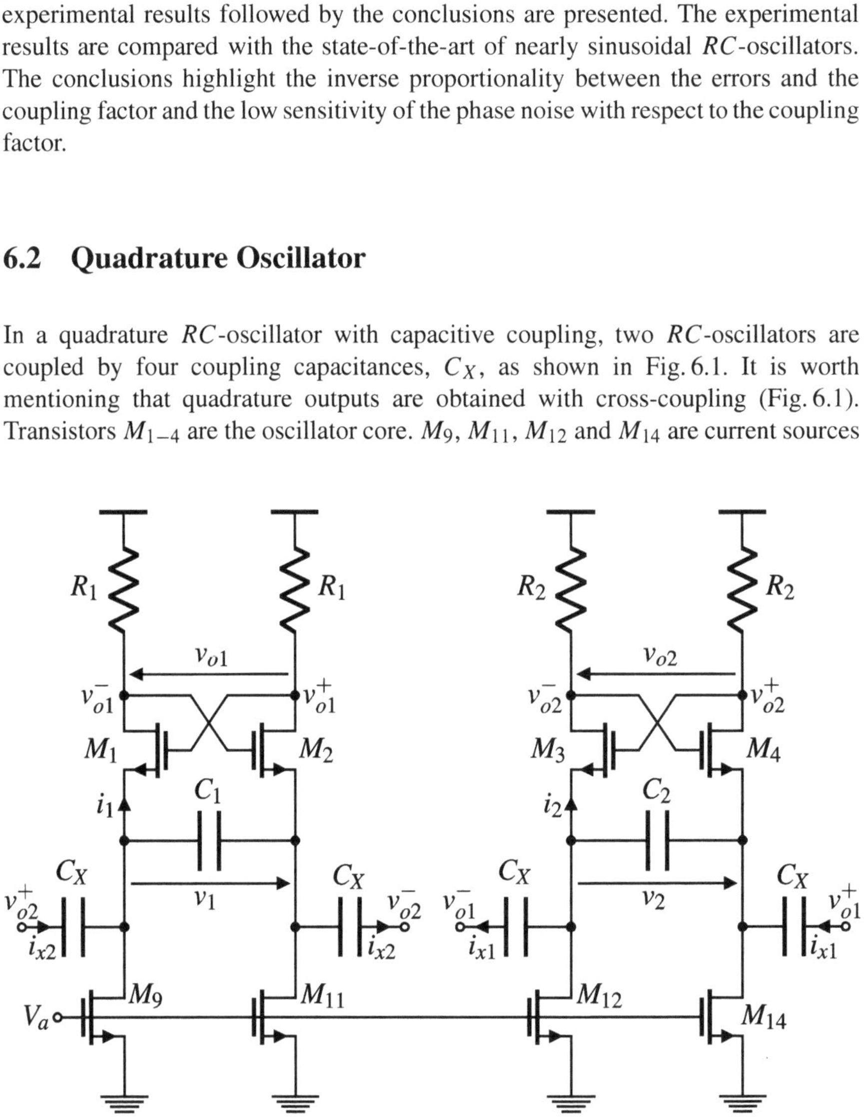

Fig. 6.1 Quadrature oscillator with capacitive coupling circuit

are parts of a multiple output current mirror. The other elements, capacitances C_1, C_2, and resistance R_1, R_2, set the amplitude and oscillation frequency.

The incremental circuit of this quadrature oscillator is obtained by substituting each individual oscillator by the series VDPO and substituting the capacitive coupling network by two-port networks. The first step, as it was shown in Chap. 5, is a valid approximation. The analysis given below shows that the second step is also a valid approximation.

At the end of this section, to validate the theoretical analysis, we present the design of a 2.4-GHz oscillator and provide the simulation results.

6.2.1 Two-Port Modeling of Capacitive Coupling Networks

Modeling the capacitive coupling as a two port network simplifies the analysis of the quadrature oscillator, since the oscillator can be reduced to two driven VDPOs (similar to the one shown in Sect. 4.3.2). However, it is worth noting that a passive network cannot guarantee the port condition (i.e. the antisymmetric currents flowing into the two terminals of each port [9]). In passive networks, the current flowing into each terminal is dependent on the external circuits connected to the network (in this case, two RC-oscillators). The port condition requires that these circuits cannot inject common-mode currents. This is only possible for ideal current sources without circuit mismatches. To better understand this requirement, consider the circuit of Fig. 6.2, where each port of the capacitive network is connected to differential- and common-mode voltage sources. From the circuit of Fig. 6.2, we can obtain the equations of the currents that flow into each terminal:

Fig. 6.2 Capacitive network

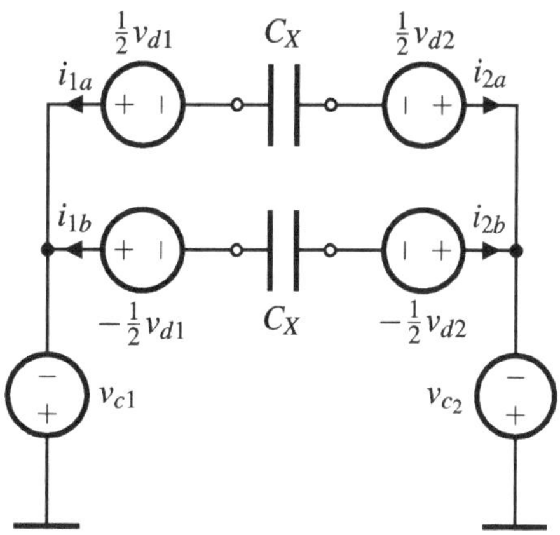

$$\begin{cases} I_{1_a} = -I_{2_a} = sC_X\left(V_{c_1} - V_{c_2}\right) + s\dfrac{C_X}{2}\left(V_{d_1} - V_{d_2}\right) & (6.1a) \\ I_{1_b} = -I_{2_b} = sC_X\left(V_{c_1} - V_{c_2}\right) - s\dfrac{C_X}{2}\left(V_{d_1} - V_{d_2}\right). & (6.1b) \end{cases}$$

Adding the equations in Eq. (6.1) eliminates the term dependent on the differential voltages (second term on the right-hand side of both equations) and doubles the term dependent on the common-mode voltages, resulting in:

$$I_{1a} + I_{1b} = -\left(I_{2a} + I_{2b}\right) = s2C_X\left(V_{c1} - V_{c2}\right). \tag{6.2}$$

If the common-mode voltages V_{c1} and V_{c2} are equal, the port condition is met ($I_{1a} + I_{1b} = -I_{2a} - I_{2b} = 0$). Thus, for this condition, the capacitive coupling network can be modeled by a two-port network.

From the small-signal circuit (Fig. 6.3a), it should be noted that resistance R_1 and R_2 are grounded. If resistors are matched there is no common mode and the capacitive coupling can be modeled by a two-port network. Since mismatches exist in practical circuits and the dynamic resistances of the current sources are not infinite, the two-port modeling is an approximation. Moreover, even without mismatches,

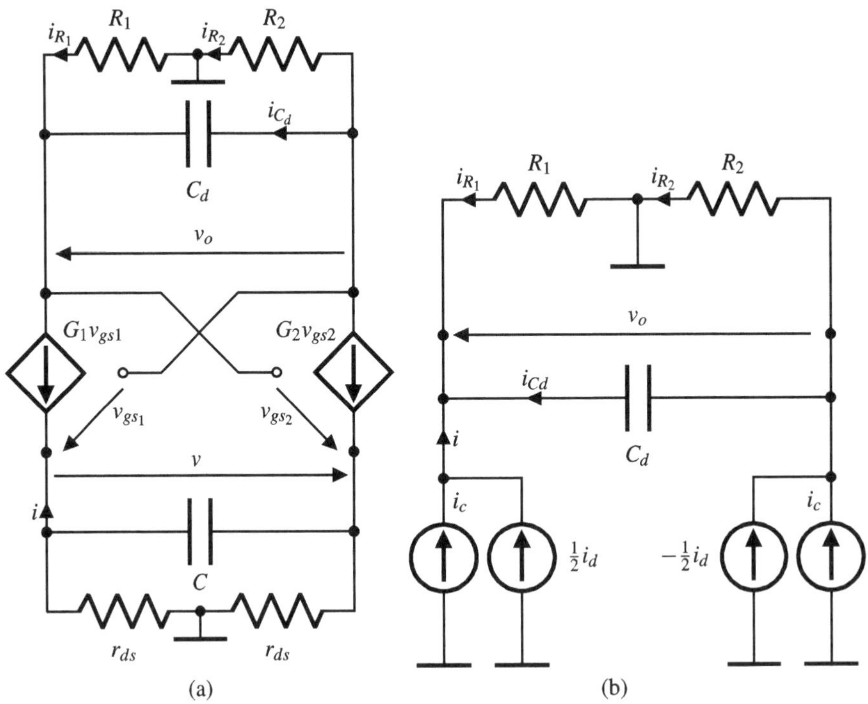

Fig. 6.3 Incremental circuit of a single *RC*-oscillator (**a**) and simplified partial circuit (**b**)

different common-mode voltages should be expected, since the coupling network is connected to different nodes in each oscillator. It is assumed here that the current sources have high dynamic resistances and that the mismatches are small, below 1%.

We simplify the schematic of the single RC-oscillator (Fig. 6.3a) as shown in Fig. 6.3b. Applying the KCL, KVL and to the circuit of Fig. 6.3b, and using the Laplace transforms one obtains

$$\begin{cases} \frac{1}{2}I_d + I_c + I_{C_d} + I_{R_1} = 0 & (6.3a) \\ I_{C_d} = sC_d V_o & (6.3b) \\ V_o = R_2 I_{R_2} + R_1 I_{R_1}. & (6.3c) \end{cases}$$

Substituting Eq. (6.3a) and Eq. (6.3b) into Eq. (6.3c), we obtain the voltage V_o as a function of both the differential- and common-mode currents:

$$V_o = \frac{R_2 - R_1}{1 + s(R_1 + R_2)C_d} I_c - \frac{1}{2}\frac{R_2 + R_1}{1 + s(R_1 + R_2)C_d} I_d, \tag{6.4}$$

from which we can see that V_o depends on the common-mode current, I_c. If there is no mismatch between R_1 and R_2 (i.e. $R_1 = R_2 = R$), the common-mode term can be omitted. A similar conclusion can be drawn for the voltage V with respect to the dynamic resistances, r_{ds}. If there are no mismatches in the oscillators, which eliminates the common-mode voltages, the capacitive coupling network can be modeled as a two-port network. However, since the common-mode terms are proportional to the mismatch, as can be seen in the first term on the right-hand side of Eq. (6.4), we will assume small mismatches and neglect the common-mode terms to use the two-port model for the case with mismatches.

With the above assumptions, the coupling networks can be substituted by the two-port equivalent circuits, as shown in Fig. 6.4. To model the capacitive coupling network, we use the admittance parameters (y-parameters) to express the terminal currents as dependent variables controlled by the ports' voltages. The admittance matrix is

$$\begin{bmatrix} y_{11} & y_{12} \\ y_{21} & y_{22} \end{bmatrix} = \begin{bmatrix} s\frac{C_X}{2} & -s\frac{C_X}{2} \\ -s\frac{C_X}{2} & s\frac{C_X}{2} \end{bmatrix}. \tag{6.5}$$

Note that the controlled current sources at the bottom two port in comparison to those at the top two port (Fig. 6.4) are antisymmetric due to the cross-coupling. The input and output impedances, y_{11}^{-1} and y_{22}^{-1}, do not have real part and, to simplify the result, are added to C_d and C, respectively, increasing both by $C_X/2$. Thus, the new capacitances of the ith oscillator are $C'_{di} = C_d + C_X/2$ and $C'_i = C_i + C_X/2$.

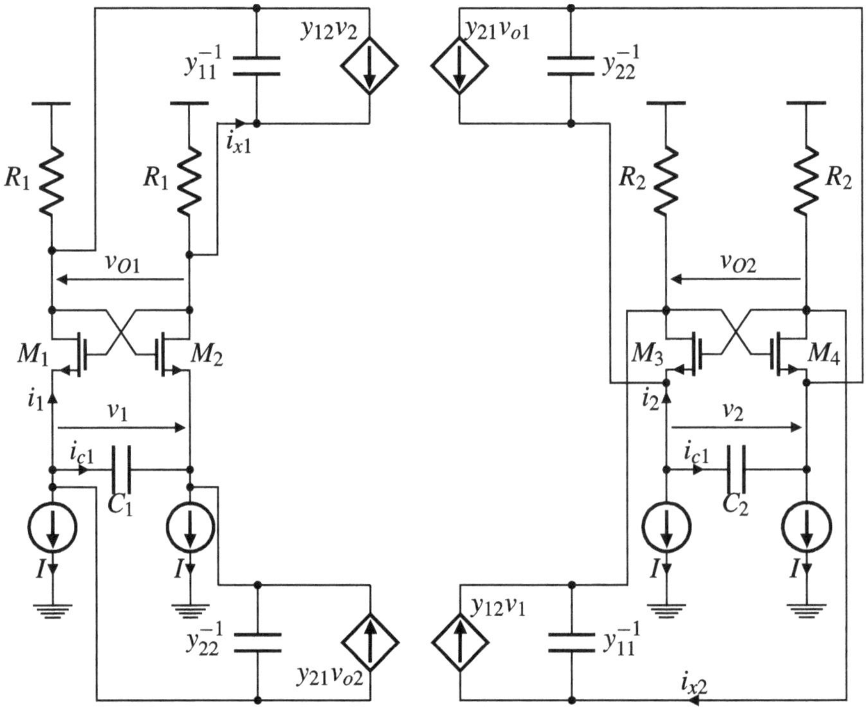

Fig. 6.4 Coupling with two-port networks

6.2.2 Incremental Model

Substituting each oscillator by the equivalent VDPO model of Fig. 4.1, in Chap. 4, results in the circuit of Fig. 6.5.

Each oscillator is driven by two coupling currents, as shown in Fig. 6.5. By applying the KVL to the circuit on the left side of Fig. 6.5, we obtain

$$L_1\frac{di_1}{dt} + R_{NL1}i_1 + R_{o1}(i_1 - i_{N1}) + R_{N1}(i_1 - i_{N1}) + R_{N1}\frac{C_X}{2}\frac{dv_2}{dt} - \frac{1}{C_1'}\int i_1 dt - \frac{1}{C_1'}\int\left[\frac{C_X}{2}\frac{dv_{o2}}{dt}\right]dt = 0, \tag{6.6}$$

where i_{N1} has a nonlinear term given by:

$$i_{N1} = K^2 i_1^3. \tag{6.7}$$

The current i_{N1} models the nonlinearities of the oscillators' core transistors. Rearranging the terms in Eq. (6.6) leads to

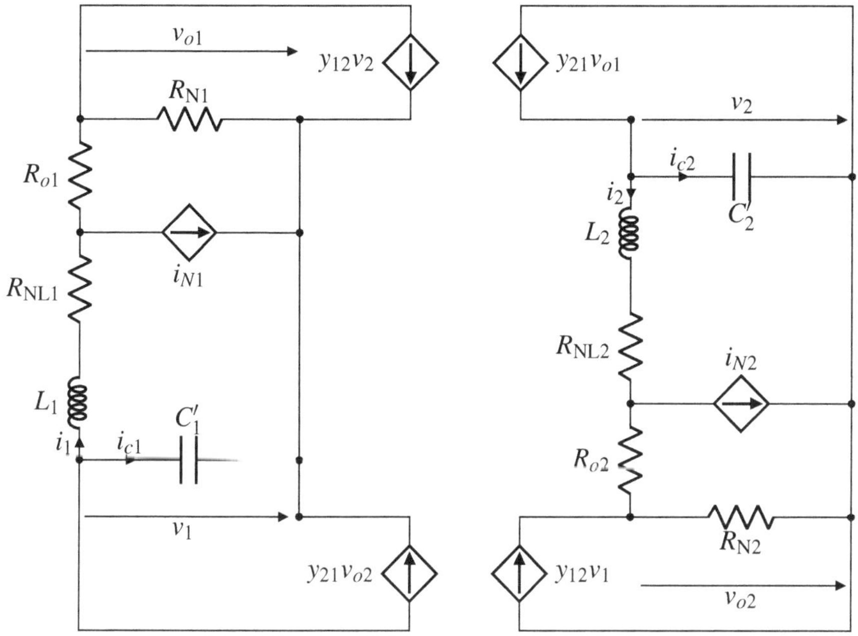

Fig. 6.5 Coupled VDPOs

$$\frac{di_1}{dt} + \frac{R_{NL1} + R_{o1} + R_{N1}}{L_1} i_1 - \frac{R_{o1} + R_{N1}}{L_1} i_{N1} + \frac{1}{L_1 C_1'} \int i_1 dt$$
$$= -R_{N1} \frac{C_X}{2} \frac{dv_2}{dt} + \frac{C_X}{2C_1'} v_{o2}. \tag{6.8}$$

From the incremental model of Fig. 6.5:

$$\begin{cases} \frac{dv_1}{dt} = -\frac{i_1}{C_1'} + \frac{C_X}{2} \frac{dv_{o2}}{dt}, & (6.9a) \\ \frac{dv_2}{dt} = -\frac{i_2}{C_2'} - \frac{C_X}{2} \frac{dv_{o1}}{dt}, & (6.9b) \\ v_{o1} = R_{N1} (i_1 - i_{N1}) + R_{N1} \frac{C_X}{2} \frac{dv_2}{dt}, & (6.9c) \\ v_{o2} = R_{N2} (i_2 - i_{N2}) - R_{N2} \frac{C_X}{2} \frac{dv_1}{dt}. & (6.9d) \end{cases}$$

The output voltages are obtained by substituting Eq. (6.9b) into Eq. (6.9c) and Eq. (6.9a) into Eq. (6.9d) resulting in:

$$\begin{cases} v_{o1} = R_{N1}\left(i_1 - i_{N1}\right) - R_{N1}\dfrac{C_X}{2C_2'}i_2 - R_{N1}\dfrac{C_X^2}{4}\dfrac{dv_{o1}}{dt}, & (6.10a) \\ v_{o2} = R_{N2}\left(i_2 - i_{N2}\right) - R_{N2}\dfrac{C_X}{2C_1'}i_1 - R_{N2}\dfrac{C_X^2}{4}\dfrac{dv_{o2}}{dt}. & (6.10b) \end{cases}$$

The third terms on the right-hand side of both equations, in Eq. (6.10), are small in comparison with the other terms and, therefore, they are neglected in the analysis that follows. Moreover, currents $i_{Ni}(i = 1, 2)$ are small in comparison with the oscillator current (i.e. $i_{Ni} \ll i_i$). Neglecting these terms as well, Eqs. (6.10) are reduced to:

$$\begin{cases} v_{o1} \approx R_{N1}i_1 - R_{N1}\alpha_2 i_2, & (6.11a) \\ v_{o2} \approx R_{N2}i_2 + R_{N2}\alpha_1 i_1, & (6.11b) \end{cases}$$

where α_i is the coupling factor:

$$\alpha_i = \frac{C_X}{2C_i + C_X} \qquad i = 1, 2.$$

The derivatives of the input voltages are obtained by substituting Eq. (6.11a) and Eq. (6.11b) into Eq. (6.9a) and Eq. (6.9b), respectively, resulting in both v_1 and v_2 as functions of the oscillators' currents:

$$\begin{cases} \dfrac{dv_1}{dt} = -\dfrac{i_1}{C_1'} + R_{N2}\dfrac{C_X}{2}\dfrac{di_2}{dt} + R_{N2}\alpha_1\dfrac{C_X}{2}\dfrac{di_1}{dt}, & (6.12a) \\ \dfrac{dv_2}{dt} = -\dfrac{i_2}{C_2'} - R_{N1}\dfrac{C_X}{2}\dfrac{di_1}{dt} + R_{N1}\alpha_2\dfrac{C_X}{2}\dfrac{di_2}{dt}, & (6.12b) \end{cases}$$

Substituting Eq. (6.11b) and Eq. (6.12b) into Eq. (6.8) results in:

$$L_1\frac{di_1}{dt} + \left(R_{NL1} + R_{o1} + R_{N1}\right)i_1 - \left(R_{o1} + R_{N1}\right)i_{N1} + \frac{1}{C_1'}\int i_1 dt$$
$$= R_{N1}\alpha_2 i_2 + R_{N1}^2\frac{C_X^2}{4}\frac{di_1}{dt} - R_{N1}^2\frac{C_X^2}{4}\alpha_2\frac{di_2}{dt} + \alpha_1 R_{N2}i_2 + R_{N2}\alpha_1^2 i_1. \tag{6.13}$$

The terms with C_X^2 and α_1^2 parameters on the right-hand side of Eq. (6.13) can be neglected. Thus, neglecting these terms, dividing both sides of Eq. (6.13) by L_1, and differentiating, we obtain

$$\frac{d^2i_1}{dt^2} + \frac{R_{NL1} + R_{o1} + R_{N1}}{L_1}\frac{di_1}{dt} - \frac{3K^2\left(R_{o1} + R_{N1}\right)i_1^2}{L_1}\frac{di_1}{dt} + \omega_{01}^2\left(1 - \alpha_1\right)i_1$$
$$\approx \frac{R_{N1}\alpha_2 + R_{N2}\alpha_1}{L_1}\frac{di_2}{dt}. \tag{6.14}$$

It is interesting to analyse Eq. (6.14), where the right-hand side is the sum of the coupling currents injected by the second oscillator into the first. Equation (6.14) corresponds to a driven VDPO. Writing Eq. (6.14) in the van der Pol's form results in:

$$\frac{\mathrm{d}^2 i_1}{\mathrm{d}t^2} - 2\left(\delta_0 - \delta_2 i_1^2\right)\frac{\mathrm{d}i_1}{\mathrm{d}t} + \omega_{01}^2\left(1 - \alpha_1\right) i_1$$
$$\approx -\left(\alpha_1 R_2 + \alpha_2 R_1\right)\frac{\omega_{01}}{Q_1 R_1}\frac{\mathrm{d}i_2}{\mathrm{d}t}, \tag{6.15}$$

where the VDP parameters are given by:

$$\delta_0 = \frac{R_{\mathrm{NL1}} + R_{\mathrm{o1}} + R_{\mathrm{N1}}}{L_1} = \frac{R_1\left(1 - \frac{C_d}{C}\right) - g_{m0}{}^{-1}}{L_1}, \tag{6.16}$$

and

$$\delta_2 = -\frac{3K^2\left(R_{\mathrm{o1}} + R_{\mathrm{N1}}\right)}{L_1} = \frac{3K^2 R_1\left(1 - \frac{C_d}{C}\right)}{L_1}. \tag{6.17}$$

For the second oscillator with the same simplification

$$\frac{\mathrm{d}^2 i_2}{\mathrm{d}t^2} - 2\left(\gamma_0 - \gamma_2 i_1^2\right)\frac{\mathrm{d}i_2}{\mathrm{d}t} + \omega_{02}^2\left(1 - \alpha_2\right) i_2 \approx \left(\alpha_2 R_1 + \alpha_1 R_2\right)\frac{\omega_{02}}{Q_2 R_2}\frac{\mathrm{d}i_1}{\mathrm{d}t}, \tag{6.18}$$

where the VDP parameters are given by:

$$\gamma_0 = \frac{R_{\mathrm{NL2}} + R_{\mathrm{o2}} + R_{\mathrm{N2}}}{L_1} = \frac{R_2\left(1 - \frac{C_d}{C}\right) - g_{m0}{}^{-1}}{L_2}, \tag{6.19}$$

$$\gamma_2 = -\frac{3K^2\left(R_{\mathrm{o2}} + R_{\mathrm{N2}}\right)}{L_1} = \frac{3K^2 R_2\left(1 - \frac{C_d}{C}\right)}{L_2}. \tag{6.20}$$

From the third term on the left-hand side of Eqs. (6.15) and (6.18), we see that the frequency should decrease when the coupling factor, α, increases. This is consistent with the intuitive idea that increasing the capacitance lowers the frequency. However, as we show next, the forcing term (the right-hand side of Eqs. (6.15) and (6.18)) opposes to this tendency and forces the oscillation frequency to increase.

To solve the differential equations, Eqs. (6.15) and (6.18), we use the harmonic balance method [10], with the assumptions of a slow varying amplitude and phase, and neglecting the high-order terms. Thus, the solutions have the form:

$$i_1(t) = I_{o1}(t)\sin(\omega t - \phi_1),$$
$$i_2(t) = I_{o2}(t)\sin(\omega t - \phi_2). \tag{6.21}$$

where I_{oi} is the current amplitude, ϕ_i the phase and ω is the frequency of oscillation. We assume that both oscillators work at the same frequency, but with different phases. The harmonic balance method simplifies the problem, by considering the amplitudes' envelopes and phases. Two nonlinear second-order differential equations are reduced to the following system of four differential equations of the first order:

$$\begin{cases} \dfrac{dI_{o1}}{dt} = \left(\delta_0 I_{o1} - \dfrac{1}{4}\delta_2 I_{o1}^3\right) + \dfrac{\alpha_{K1}}{2} I_{o2}\cos(\Delta\phi) & \text{(6.22a)} \\ \dfrac{dI_{o2}}{dt} = \left(\gamma_0 I_{o2} - \dfrac{1}{4}\gamma_2 I_{o2}^3\right) + \dfrac{\alpha_{K2}}{2} I_{o1}\cos(\Delta\phi) & \text{(6.22b)} \\ \dfrac{d\phi_1}{dt} = \dfrac{\omega^2 - \omega_{01}^2(1-\alpha_1)}{2\omega} + \dfrac{\alpha_{K1}}{2}\dfrac{I_{o2}}{I_{o1}}\sin(\Delta\phi) & \text{(6.22c)} \\ \dfrac{d\phi_2}{dt} = \dfrac{\omega^2 - \omega_{02}^2(1-\alpha_2)}{2\omega} - \dfrac{\alpha_{K2}}{2}\dfrac{I_{o1}}{I_{o2}}\sin(\Delta\phi), & \text{(6.22d)} \end{cases}$$

where α_{K1} and α_{K2} are

$$\alpha_{K1} = -(\alpha_1 R_2 + \alpha_2 R_1)\frac{\omega_{01}}{Q_1 R_1}; \quad \alpha_{K2} = (\alpha_2 R_1 + \alpha_1 R_2)\frac{\omega_{02}}{Q_2},$$

and $\Delta\phi = \phi_2 - \phi_1$ is the phase difference of i_1 and i_2. From Eq. (6.22), we can derive the steady-state equations for the amplitudes, frequency and phase. In the next subsection, we analyse first the oscillator without mismatches and after that the case with mismatches.

6.2.3 *Oscillators Without Mismatches*

If there are no mismatches, $R_1 = R_2 = R$, $C_1 = C_2 = C$, $L_1 = L_2 = L$ and $\alpha_1 = \alpha_2 = \alpha$. The free-running frequencies are equal, $\omega_{01} = \omega_{02} = \omega_0$, and the VDP parameters are $\delta_0 = \gamma_0$ and $\delta_2 = \gamma_2$, and the quality factors are $Q_1 = Q_2 = Q$. The coupling factors are antisymmetrical, $\alpha_{K2} = -\alpha_{K1} = \alpha_K$.

With the above assumptions, we simplify the system of differential equations (6.22) and derive the steady-state solutions (equilibrium points). This analysis shows that, without mismatches, the amplitudes are equal, the oscillators are in perfect quadrature, and the oscillation frequency increases with the coupling factor. In the following, to understand how the circuit reaches the steady state, a transient analysis is done, by linearizing the system near the equilibrium points. The transient analysis shows that the stable equilibrium points correspond to nonzero amplitudes.

Before we proceed, it is important to define the equilibrium points notation and their coordinates. Although the system equation (6.22) has four equations, Eqs. (6.22c) and (6.22d) can be merged, if we use the phase difference, $\Delta\phi$, instead of the phase of each oscillator. Hence, a three-dimensional coordinate system that we refer to as phase space can be used. We denote an equilibrium point by E with a subscript. The position of each equilibrium point is uniquely determined by three coordinates (in the phase space), identified by a triplet, where the first and second coordinates represent, respectively, the amplitude I_{o1} and I_{o2}, and the third coordinate is the phase difference, $\Delta\phi$.

The equilibrium points, of the system of differential equations (6.22), are obtained by setting all derivative terms equal to zero (i.e. $\frac{dI_{o1}}{dt} = \frac{dI_{o1}}{dt} = 0$ and $\frac{d\phi_1}{dt} = \frac{d\phi_2}{dt} = 0$). To avoid the indeterminate form $\frac{0}{0}$, we multiply Eq. (6.22c) by I_{o1} and Eq. (6.22d) by I_{o2} reducing Eq. (6.22) to:

$$\begin{cases} \left(\delta_0 I_{o1} - \frac{1}{4}\delta_2 I_{o1}^3\right) + \frac{\alpha_K}{2} I_{o2} \cos(\Delta\phi) = 0 & (6.23a) \\ \left(\delta_0 I_{o2} - \frac{1}{4}\delta_2 I_{o2}^3\right) + \frac{\alpha_K}{2} I_{o1} \cos(\Delta\phi) = 0 & (6.23b) \\ \frac{\omega^2 - \omega_0^2(1-\alpha)}{2\omega} I_{o1} + \frac{\alpha_K}{2} I_{o2} \sin(\Delta\phi) = 0 & (6.23c) \\ \frac{\omega^2 - \omega_0^2(1-\alpha)}{2\omega} I_{o2} - \frac{\alpha_K}{2} I_{o1} \sin(\Delta\phi) = 0. & (6.23d) \end{cases}$$

From Eq. (6.23), we find equilibrium points where $I_{o1} = I_{o2} = 0$, meaning that the oscillators do not start. Thus, in the phase space on the $\Delta\phi$ axis we have an infinite number of equilibrium points that we represent by $E_0 = (0, 0, \Delta\phi)$. Luckily, these equilibrium points are not stable and, therefore, the circuit thermal noise guarantees that the oscillators start.

Four more equilibrium points exist for Eq. (6.23), if we consider negative amplitudes and quadrature outputs $\Delta\phi = \frac{\pi}{2}$ and $\Delta\phi = -\frac{\pi}{2}$. Since physically these four solutions are the same, we consider only the equilibrium point of the first quadrant, with positive amplitudes and positive phase difference. Thus, at equal amplitudes $I_{o1} = I_{o2} = I_{osc}$ and in quadrature, $\Delta\phi = \frac{\pi}{2}$, we have the equilibrium point $E_1 = (I_{osc}, I_{osc}, \frac{\pi}{2})$. Then, the current amplitude, I_{osc}, is obtained by solving Eq. (6.23a) or Eq. (6.23b) with respect to I_{osc}, with a $\Delta\phi = \frac{\pi}{2}$, resulting in:

$$I_{osc} = 2\sqrt{\frac{\delta_0}{\delta_2}} = 8I\sqrt{\frac{Rg_{m0} - \left(\frac{C}{C-C_d}\right)}{3Rg_{m0}}}. \tag{6.24}$$

To obtain the output voltage, which is easier to compare with measurement results, we multiply Eq. (6.24) by the output resistance $2R$ and obtain the output voltage:

$$V_{\mathrm{osc}} = 16RI\sqrt{\frac{Rg_{m0} - \left(\frac{C}{C-C_d}\right)}{3Rg_{m0}}}. \tag{6.25}$$

The oscillation frequency, ω, is derived by combining Eq. (6.23c) and Eq. (6.23d), note that for equal amplitudes $I_{o1} = I_{o2} = I_{\mathrm{osc}}$, resulting in:

$$\omega^2 - \omega\alpha_K \sin(\Delta\phi) - \omega_0^2(1-\alpha) = 0. \tag{6.26}$$

Two cases should be considered for Eq. (6.26): the first is for $\Delta\phi = \pi/2$ and the second for $\Delta\phi = -\pi/2$. Each equation has two solutions leading to a total of four solutions. However, the negative frequency solutions can be ruled out, and two possible solutions remain. Thus, the positive frequency solutions of Eq. (6.26) for the two phase shifts are

$$\begin{cases} \omega = \dfrac{2\alpha\omega_0}{2Q} + \dfrac{1}{2}\sqrt{\dfrac{4\alpha^2\omega_0^2}{Q^2} + 4\omega_0^2(1-\alpha)}, & \text{if } \Delta\phi = \frac{\pi}{2} \quad (6.27\text{a}) \\ \omega = -\dfrac{2\alpha\omega_0}{2Q} + \dfrac{1}{2}\sqrt{\dfrac{4\alpha^2\omega_0^2}{Q^2} + 4\omega_0^2(1-\alpha)}, & \text{if } \Delta\phi = -\frac{\pi}{2}. \quad (6.27\text{b}) \end{cases}$$

Using the approximation $\sqrt{1 \pm x} \approx (1 \pm x/2)$ for $|x| \ll 1$, one can simplify the result to the following:

$$\begin{cases} \omega \approx \omega_0\left(1 + \dfrac{2 - Q + 2\alpha/Q}{2Q}\alpha\right), & \text{if } \Delta\phi = \frac{\pi}{2} \quad (6.28\text{a}) \\ \omega \approx \omega_0\left(1 - \dfrac{2 + Q - 2\alpha/Q}{2Q}\alpha\right). & \text{if } \Delta\phi = -\frac{\pi}{2}. \quad (6.28\text{b}) \end{cases}$$

The system Eq. (6.28) shows that the oscillator can operate in one of two modes. If $Q \approx 1$ (RC-oscillator), then the mode frequencies are

$$\begin{cases} \omega \sim \omega_0(1 + 0.5\alpha), & \text{if } \Delta\phi = \frac{\pi}{2} \quad (6.29\text{a}) \\ \omega \sim \omega_0(1 - 1.5\alpha), & \text{if } \Delta\phi = -\frac{\pi}{2}. \quad (6.29\text{b}) \end{cases}$$

The results in Eq. (6.29) are interesting. In the first mode, $\Delta\phi = \pi/2$, the increase of the capacitances is counteracted by the coupling mechanism, in such a way that the frequency increases, as shown in Eq. (6.29a). However, for the second mode, $\Delta\phi = -\frac{\pi}{2}$, when the coupling factor, α, increases the oscillation frequency decreases. This result makes physical sense, since the circuit capacitances increased. The coupling adds to the capacitive load decreasing further the frequency with the increase of the coupling factor; this also explains the asymmetry between the two modes. Moreover, as we will show at the end of this chapter, both modes are stable, mutually exclusive and both are possible in practice. This situation, called bimodal oscillation, has been already identified in coupled LC-oscillators. Although, both modes are stable, in practice, with proper initial conditions, the prevailing mode can be selected [7]. The analysis of the second mode has little interest. Thus, we focus the research on the first mode with $\Delta\phi = \pi/2$.

6.2.4 Stability of the Equilibrium Points

To understand how the circuit reaches the steady state, we perform the transient analysis by deriving the phase-space paths. Simplifying the system equation (6.22) for antisymmetrical coupling factors, $\alpha_{K1} = -\alpha_{K2} = \alpha_K$, and combining Eqs. (6.22c) and (6.22d), we obtain

$$\begin{cases} \dfrac{\mathrm{d}I_{o1}}{\mathrm{d}t} = \delta_0 I_{o1} - \dfrac{1}{4}\delta_2 I_{o1}^3 - \dfrac{\alpha_K}{2} I_{o2}\cos\Delta\phi & (6.30a) \\ \dfrac{\mathrm{d}I_{o2}}{\mathrm{d}t} = \delta_0 I_{o2} - \dfrac{1}{4}\delta_2 I_{o2}^3 + \dfrac{\alpha_K}{2} I_{o1}\cos\Delta\phi. & (6.30b) \\ \dfrac{\mathrm{d}\Delta\phi}{\mathrm{d}t} = -\dfrac{\alpha_K}{2}\left(\dfrac{I_{o1}}{I_{o2}} - \dfrac{I_{o2}}{I_{o1}}\right)\sin\Delta\phi. & (6.30c) \end{cases}$$

Although these are first-order differential equations, they are nonlinear, which leads to a cumbersome analysis. For convenience, we analyse the transient in the vicinity of the equilibrium point, where the system can be linearized. Thus, by linearizing the system Eq. (6.30) we obtain

$$\begin{cases} \dfrac{\mathrm{d}I_{o1}}{\mathrm{d}t} \approx \left(\delta_0 - \dfrac{3}{4}\delta_2 I_E^2\right)(I_{o1} - I_E) - \dfrac{\alpha_K}{2}\cos\Delta\phi_E\,(I_{o2} - I_E) & \\ +\dfrac{\alpha_K}{2} I_E\,(\Delta\phi - \Delta\phi_E) & (6.31a) \\ \dfrac{\mathrm{d}I_{o2}}{\mathrm{d}t} \approx \left(\delta_0 - \dfrac{3}{4}\delta_2 I_E^2\right)(I_{o2} - I_E) + \dfrac{\alpha_K}{2}\cos\Delta\phi_E\,(I_{o1} - I_E) & \\ -\dfrac{\alpha_K}{2} I_E\,(\Delta\phi - \Delta\phi_E), & (6.31b) \\ \dfrac{\mathrm{d}\Delta\phi}{\mathrm{d}t} \approx \dfrac{\alpha_K}{I_E}(I_{o2} - I_{o1}). & (6.31c) \end{cases}$$

where I_E and $\Delta\phi_E$ are, respectively, the steady-state current and the phase difference at the equilibrium points.

At the equilibrium point E_0, the current amplitudes are equal and, therefore, Eq. (6.31c) is equal to zero and the system Eq. (6.31) is reduced to two equations that can be written in matrix form as:

$$\begin{bmatrix} \frac{dI_{o1}}{dt} \\ \frac{dI_{o2}}{dt} \end{bmatrix} = J \cdot \begin{bmatrix} I_{o1} \\ I_{o2} \end{bmatrix},$$

where J is the Jacobian matrix. The Jacobian matrix at E_0 is given by:

$$J_{E_0} = \begin{bmatrix} \frac{\partial}{\partial I_{o1}}\left(\frac{dI_{o1}}{dt}\right) & \frac{\partial}{\partial I_{o1}}\left(\frac{dI_{o2}}{dt}\right) \\ \frac{\partial}{\partial I_{o2}}\left(\frac{dI_{o1}}{dt}\right) & \frac{\partial}{\partial I_{o2}}\left(\frac{dI_{o2}}{dt}\right) \end{bmatrix} = \begin{bmatrix} \delta_0 & \frac{\alpha_K}{2}\cos\Delta\phi_E \\ -\frac{\alpha_K}{2}\cos\Delta\phi_E & \delta_0 \end{bmatrix}.$$

To check the stability of an equilibrium point, we must determine the characteristic equation:

$$\lambda^2 - T\lambda + D = 0, \tag{6.32}$$

where T is the trace of J (sum of the main diagonal elements) and D is the determinant of J. The conditions for stability are $T < 0$ and $D > 0$ [10]. For $T > 0$ or $D < 0$, the equilibrium point is unstable [10]. From the Jacobian matrix J_{E_0} at the equilibrium point E_0, we obtain the characteristic equation:

$$\lambda^2 - 2\delta_0\lambda + \left(\delta_0^2 + \frac{\alpha_K^2}{4}(\cos\Delta\phi)^2\right) = 0. \tag{6.33}$$

From Eq. (6.33), if $\delta_0 > 0$ we conclude that all the equilibrium points on the $\Delta\phi$ axis are unstable, because both the trace and the determinant are positive. However, we have two different cases. When $\Delta\phi = \pi/2$, the eigenvalues are real. This means that near that point we have an unstable node. For $\Delta\phi \neq \pi/2$, the eigenvalues are complex conjugate with a negative real part which means that at these points we have spiral sources. For $\Delta\phi > \pi/2$, the spiral direction is counterclockwise, and for $\Delta\phi < \pi/2$ it is clockwise. The relevant conclusion is that the points near the origin or along the $\Delta\phi$-axis are unstable, which means that the oscillator will start.

Another equilibrium point exists at $E_1 = (I_{osc}, I_{osc}, \pi/2)$. However, it should be noted that $\sin\Delta\phi \neq 0$ on the right-hand side of Eq. (6.30) is making the problem three dimensional, as shown in Fig. 6.6.

The system Eq. (6.31) at E_1 becomes

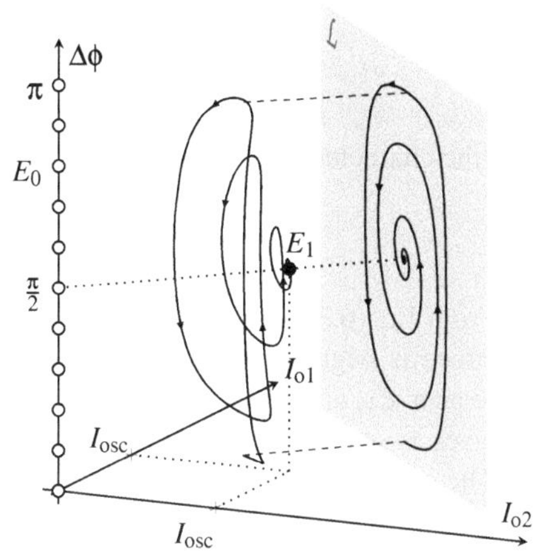

Fig. 6.6 Phase space of the capacitive coupling oscillator

$$\begin{cases} \dfrac{dI_{o1}}{dt} \approx -2\delta_0 (I_{o1} - I_E) + \dfrac{\alpha_K}{2} I_E \left(\Delta\phi - \dfrac{\pi}{2}\right) & (6.34a) \\ \dfrac{dI_{o2}}{dt} \approx -2\delta_0 (I_{o2} - I_E) - \dfrac{\alpha_K}{2} I_E \left(\Delta\phi - \dfrac{\pi}{2}\right) & (6.34b) \\ \dfrac{d\Delta\phi}{dt} \approx \dfrac{\alpha_K}{I_E} (I_{o2} - I_{o1}) . & (6.34c) \end{cases}$$

Near E_1, we can reduce the system to two dimensions by projecting the paths onto the plane $\mathcal{L}$, as illustrated by Fig. 6.6. The plane $\mathcal{L}$ can be any plane perpendicular to the straight line containing the points $(0, 0, \pi/2)$ and E_1. If such a plane is chosen, we can make the simple transformations:

$$\begin{cases} \Delta I = I_{o2} - I_{o1} \\ \phi_\epsilon = \Delta\phi - \frac{\pi}{2}, \end{cases} \qquad (6.35)$$

where ΔI is the amplitude error and ϕ_ϵ is the phase error. Then, subtracting Eq. (6.34a) from Eq. (6.34b), the system can be reduced to:

$$\begin{cases} \dfrac{d\Delta I}{dt} \approx -2\delta_0 (\Delta I) + \alpha_K I_{osc} \phi_\epsilon & (6.36a) \\ \dfrac{d\phi_\epsilon}{dt} \approx \dfrac{\alpha_K}{I_{osc}} \Delta I & (6.36b) \end{cases}$$

The equilibrium point E_1 in the new coordinates is located at the origin $\widehat{E}_1 = (\Delta I = 0, \phi_\epsilon = 0)$. Hence, the Jacobian matrix is given by:

$$J_{\widehat{E}_1} = \begin{bmatrix} -2\delta_0 & -\alpha_K I_{\text{osc}} \\ \frac{\alpha_K}{I_{\text{osc}}} & 0 \end{bmatrix}, \tag{6.37}$$

and the characteristic equation is

$$\lambda^2 + 2\delta_0\lambda + \alpha_K^2 = 0. \tag{6.38}$$

From Eq. (6.38), we can conclude that the equilibrium point $\widehat{E}_1$ is stable, since the trace is negative and the determinant is positive. If $\delta_0 > |\alpha_K|/2$, we have a node near $\widehat{E}_1$, and if $\delta_0 < |\alpha_K|/2$ it is a spiral sink. For the former case, the system behaves as an overdamped system and for the latter case as an underdamped system.

The phase portrait (in plane $\mathcal{L}$) for the second case (which occurs for low coupling factors) is shown in Fig. 6.7a. The highlighted path $\mathcal{P}$ in the phase portrait of Fig. 6.7a represents the transients for the case where both oscillators start with the same amplitude and phase. Three instants are marked along the path for easier correspondence with the time solution shown in Fig. 6.7b on the right. From path $\mathcal{L}$ and the top and middle plots on the right-hand side, we see that at instant $t = t_0$ the two oscillators are in-phase with an amplitude of 1V. Following the path direction (indicated by the arrow), we expect an increase of the amplitude error ΔI, meaning that the amplitude of the second oscillator, I_{o2}, increases and I_{o1} decreases, reaching a peak at the instant $t = t_1$. For $t > t_1$, an inversion of the trend occurs, and I_{o2} starts decreasing and I_{o1} is increasing until another inversion occurs. The cycle repeats until the equilibrium point is reached. Hence, the phase, amplitude and frequency become closer to steady-state values in each cycle, as shown in the bottom plot of Fig. 6.7b.

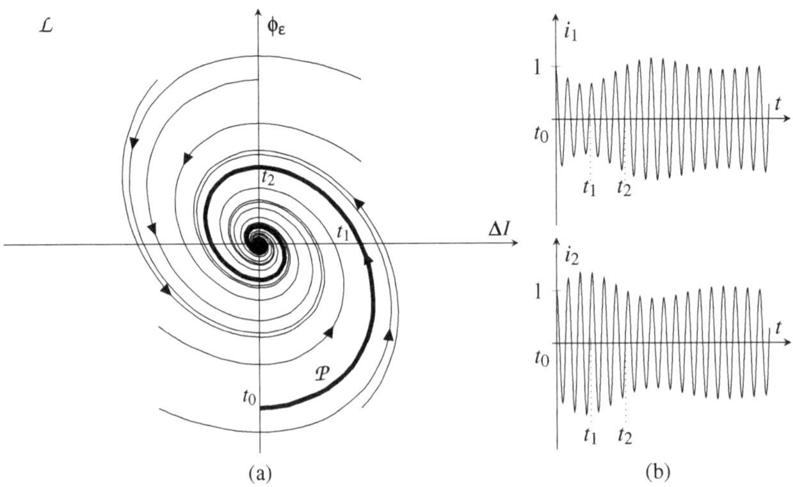

Fig. 6.7 Capacitive coupling (**a**) phase portrait and (**b**) transient for path $\mathcal{P}$

Solving Eq. (6.37), we obtain

$$\lambda = -\delta_0 \pm \sqrt{\delta_0^2 - 4\alpha_K^2}. \tag{6.39}$$

If $\delta_0 > |2\alpha_K|$, there are two negative real eigenvalues (indicating the behaviour of an overdamped system). If $\delta_0 < |2\alpha_K|$, there are two complex conjugate eigenvalues, which indicates an underdamped second-order system as shown in the phase portrait and time solution in Fig. 6.7. The amplitude envelope of the latter will have a damping factor, ζ, natural frequency, ω_n, and damped natural frequency, ω_m, respectively, given by:

$$\zeta = \frac{1}{2}(K_0R - 1); \qquad \omega_n = \frac{1}{RC}; \qquad \omega_m = \omega_n\sqrt{4|\alpha_K|^2 R^2 - (K_0R - 1)^2}. \tag{6.40}$$

For $\delta_0 < |2\alpha_K|$, the solution is given by:

$$\begin{cases} I_{o1} = I_{osc} + K_{I_{o1}} e^{-\delta_0 t}\sin(\omega_n t + \theta_v) \\ I_{o2} = I_{osc} + K_{I_{o2}} e^{-\delta_0 t}\sin(\omega_n t + \theta_v) \\ \Delta\phi = \pm\frac{\pi}{2} + K_\phi e^{-\delta_0 t}\cos(\omega_n t + \theta_\phi), \end{cases}$$

where the K_{Io1}, K_{Io2}, K_ϕ, θ_v and θ_ϕ are constants depending on the initial conditions.

During the transient, the oscillators' outputs are not in quadrature and the amplitudes are not steady. The settling time, t_s, is an important indicator to estimate the transient interval. It is defined as:

$$t_s = -\frac{\ln(\frac{\Delta I_o}{I_o})}{\zeta\omega_n} = \frac{\ln\frac{I_o}{\Delta I_o}}{\delta_0}, \tag{6.41}$$

where $\frac{\Delta I_o}{I_o}$ is the relative variation of the amplitude to consider the oscillator in steady state. Hence, the settling time for $\frac{\Delta I_o}{I_o} = 1\%$ is given by:

$$t_s > \frac{2RC\ln 100}{(K_0R - 1)} \approx \frac{9.21RC}{(K_0R - 1)}.$$

6.2.5 Mode Selection

By definition, the frequency is the derivative of the phase. Hence, from Eq. (6.21) the instantaneous frequency of the oscillator is given by:

$$\omega_i(t) = \omega_i(0) - \frac{\mathrm{d}\phi_i}{\mathrm{d}t}, \qquad i = 1, 2 \tag{6.42}$$

where $\omega_i(0)$ is the frequency at start-up ($t \approx 0$).

Let us consider that the oscillators at start-up are almost in-phase ($\phi_2 > \phi_1 \gtrsim 0$), and have low amplitudes ($I_{o1} = I_{o2} \gtrsim 0$), and their oscillation frequency starts from the free-running value, ω_0, since the coupling has negligible influence. With these conditions, Eqs. (6.22c) and (6.22d) can be approximated as:

$$\begin{cases} \dfrac{\mathrm{d}\phi_1}{\mathrm{d}t} \approx \alpha_{K_1} \cdot \Delta\phi & (6.43a) \\ \dfrac{\mathrm{d}\phi_2}{\mathrm{d}t} \approx -\alpha_{K_2} \cdot \Delta\phi, & (6.43b) \end{cases}$$

Hence, the product of the coupling factors and the phase difference determines the frequency, and the sign of the phase derivative at steady state. Then, if $\alpha_{K1} < 0$, $\alpha_{K2} > 0$ and $\phi_2 > \phi_1 \gtrsim 0$, it can be seen from Eqs. (6.22c) and (6.22d) that the derivatives of the phases are negative. This, in accordance with Eq. (6.42), leads to an increase of the oscillation frequency. If $\phi_1 > \phi_2 \gtrsim 0$, the derivatives are positive and the frequency decreases. The opposite conclusion can be drawn if we consider $\alpha_{K1} > 0$ and $\alpha_{K2} < 0$, since the derivatives will have the same sign of the phase difference.

When the frequency starts to change in one direction, it never goes back and it continues until the derivatives of the phase are zero ($\mathrm{d}\phi_1/\mathrm{d}t = \mathrm{d}\phi_2/\mathrm{d}t = 0$).

Applying the above theory to the circuit of Fig. 6.1, one concludes that if the oscillator 1 (at the left-hand side in Fig. 6.1) starts first, the high frequency mode is selected. Conversely, if the oscillator 2 (at the right-hand side in Fig. 6.1) starts first, the low-frequency mode is selected.

6.2.6 Capacitive Coupled Oscillators with Mismatches

In this section, we derive the amplitude and phase errors for the mode $\Delta\phi \approx \pi/2$, considering that there are components' mismatches. The complete derivation is cumbersome, and therefore, in this section, we present only the important steps to the final equations. In the following derivation, we assume that the oscillators' core transistors are identical, and therefore, their transconductances and capacitances are equal. We consider also that the oscillators reach the steady state, with all derivatives in Eq. (6.22) equal to zero.

We assume small mismatches (i.e. $\frac{\Delta R}{R} \ll 1\%$ and $\frac{\Delta C}{C} \ll 1\%$) and, to simplify the derivation, we neglect the terms including the product of relative mismatches (e.g. $\frac{\Delta R}{R}\frac{\Delta C}{C} \approx 0$). The error equations are derived as functions of the mismatches of R and C, considering that

$$R_1 = R\left(1 - \frac{\Delta R}{2R}\right); \qquad R_2 = R\left(1 + \frac{\Delta R}{2R}\right), \tag{6.44}$$

and

$$C_1 = C\left(1 - \frac{\Delta C}{2C}\right); \qquad C_2 = C\left(1 + \frac{\Delta C}{2C}\right). \tag{6.45}$$

The mismatches of the inductances, L_{e1} and L_{e2}, are related to the mismatches of the resistances as:

$$L_{e1} = L_e\left(1 - \frac{\Delta R}{2R}\right); \qquad L_{e2} = L_e\left(1 + \frac{\Delta R}{2R}\right), \tag{6.46}$$

where L_e is the inductance with no mismatch given by:

$$L_e = 4R{g_{m0}}^{-1}\left(C_d + C_X\right).$$

By definition, the amplitudes as functions of the amplitude error are given by:

$$I_{o1} = I_{osc}\left(1 - \frac{\epsilon_A}{2}\right); \qquad I_{o2} = I_{osc}\left(1 + \frac{\epsilon_A}{2}\right),$$

where ϵ_A is the amplitude error.

Before we derive the amplitude error equation, we need to obtain first the oscillation frequency, ω. The amplitude error is derived from Eqs. (6.22c) and (6.22d) that depend on the oscillation frequency.

The oscillation frequency is obtained by combining Eqs. (6.22c) and (6.22d). Rearranging the terms with respect to ω, substituting the resistances R_1 and R_2 given by equations in Eq. (6.44) and grouping α_1 and α_2 parameters, we obtain

$$\frac{2\omega^2 - \left[\omega_{01}^2\left(1-\alpha_1\right) + \omega_{02}^2\left(1-\alpha_2\right)\right]}{2\omega} - \left(R_1\alpha_2 + R_2\alpha_1\right)\left(\frac{1}{L_1'}\frac{I_{o2}}{I_{o1}} + \frac{1}{L_2'}\frac{I_{o1}}{I_{o2}}\right)\sin\left(\Delta\phi\right) = 0. \tag{6.47}$$

Rearranging the terms of the above equation with respect to ω, substituting the resistances R_1 and R_2 by equations, Eq. (6.44), and grouping α_1 and α_2 parameters in the second term on the left-hand side of Eq. (6.47), we obtain

$$\omega^2 - R\omega\left[\left(\alpha_1+\alpha_2\right) + \left(\frac{\Delta R}{2R}\right)\left(\alpha_1-\alpha_2\right)\right]\left(\frac{1}{L_1'}\frac{I_{o2}}{I_{o1}} + \frac{1}{L_2'}\frac{I_{o1}}{I_{o2}}\right)\sin\left(\Delta\phi\right) - \frac{1}{2}\left[\frac{1}{L_1'C_1'} + \frac{1}{L_2'C_2'}\right] = 0.$$

Using the approximations $(\alpha_1 + \alpha_2 \approx 2\alpha)$ and $\alpha_1 - \alpha_2 \approx \alpha\,(1-\alpha)\,\frac{\Delta C}{C}$ in the above equation, we obtain

$$\omega^2 - R\omega\left[2\alpha + \left(\frac{\Delta R}{2R}\right)\left(\frac{\Delta C}{C}\right)\alpha\,(1-\alpha)\right]\left(\frac{1}{L_1'}\frac{I_{o2}}{I_{o1}} + \frac{1}{L_2'}\frac{I_{o1}}{I_{o2}}\right)\sin\left(\Delta\phi\right) - \frac{1}{2}\frac{1}{LC}2\,(1-\alpha) \approx 0.$$

Assuming small mismatches (i.e. $\left(\frac{\Delta R}{R}\right) \ll 1\%$ and $\left(\frac{\Delta C}{C}\right) \ll 1\%$), so that the term $\left(\frac{\Delta R}{2R}\right)\left(\frac{\Delta C}{C}\right)(1-\alpha)$ can be neglected, we obtain

$$\omega^2 - 2R\alpha\omega\left(\frac{1}{L_1'}\frac{I_{o2}}{I_{o1}} + \frac{1}{L_2'}\frac{I_{o1}}{I_{o2}}\right)\sin\left(\Delta\phi\right) - \omega_0^2\,(1-\alpha) \approx 0. \tag{6.48}$$

Using $I_{o1}^2 \approx I_{osc}^2\,(1-\epsilon_A)$, $I_{o2}^2 \approx I_{osc}^2\,(1+\epsilon_A)$ and $I_{o1}I_{o2} \approx I_{osc}^2\left(1 - \frac{\epsilon_A^2}{4}\right) \approx I_{osc}$, the above equation can be written as:

$$\omega^2 - 2R\alpha\omega\left[\frac{L_2' - L_1'}{L_1'L_2'}\epsilon_A + \frac{L_1' + L_2'}{L_1'L_2'}\right]\sin\left(\Delta\phi\right) - \omega_0^2\,(1-\alpha) \approx 0.$$

Substituting L_{e1} and L_{e2} by their expressions of Eq. (6.46), the above equation can be rewritten as:

$$\begin{aligned}\omega^2 &- \frac{2R\alpha}{L}\omega\left[\left(1 + \frac{\Delta R}{2R} - 1 + \frac{\Delta R}{2R}\right)\epsilon_A\right.\\ &\left.+\left(1 + \frac{\Delta R}{2R} + 1 - \frac{\Delta R}{2R}\right)\right]\sin\left(\Delta\phi\right)\\ &- \omega_0^2\,(1-\alpha) \approx 0,\end{aligned} \tag{6.49}$$

which can be reduced to:

$$\omega^2 - \frac{4R\alpha}{L}\omega\sin\left(\Delta\phi\right) - \omega_0^2\,(1-\alpha) \approx 0. \tag{6.50}$$

Note that $\frac{R}{L_e} = \frac{\omega_0}{Q}$, where ω_0 is the free-running frequency. When the oscillators are synchronized and in quadrature, Eq. (6.50) can be split into two equations: one for $\Delta\phi \approx \frac{\pi}{2}$ and the other for $\Delta\phi \approx -\frac{\pi}{2}$. Thus, $\sin\left(\Delta\phi\right) \approx \pm 1$, resulting in:

$$\begin{cases} \omega^2 - \dfrac{2\omega_0}{Q}\alpha\omega - \omega_0^2(1-\alpha) \approx 0, & \text{if } \Delta\phi \approx \frac{\pi}{2} \quad (6.51a) \\ \omega^2 + \dfrac{2\omega_0}{Q}\alpha\omega - \omega_0^2(1-\alpha) \approx 0. & \text{if } \Delta\phi \approx -\frac{\pi}{2} \quad (6.51b) \end{cases}$$

Four solutions can be derived from Eq. (6.51). But, if we rule out the solution with negative frequencies, two solutions only remain. Considering $\frac{4\alpha^2}{Q^2} \ll 1$ and $\sqrt{1-x} \approx \left(1 - \frac{x}{2}\right)$ for $|x| \ll 1$ yields

$$\begin{cases} \omega \approx \omega_0 \left(1 - \dfrac{\alpha}{2} + \dfrac{\alpha}{Q} + \dfrac{\alpha^2}{2Q^2}\right) & (6.52a) \\ \omega \approx \omega_0 \left(1 - \dfrac{\alpha}{2} - \dfrac{\alpha}{Q} - \dfrac{\alpha^2}{2Q^2}\right), & (6.52b) \end{cases}$$

Assuming $Q \approx 1$, to ensure minimum phase noise, this result can be further simplified to:

$$\begin{cases} \omega \approx \omega_0 \left(1 + \dfrac{2-Q}{2Q}\alpha\right) & (6.53a) \\ \omega \approx \omega_0 \left(1 - \dfrac{2+Q}{2Q}\alpha\right). & (6.53b) \end{cases}$$

It can be seen from Eq. (6.53) that for small coupling factors the impact of the mismatches on the oscillation frequency is negligible. Moreover, equations in Eq. (6.53) are identical to those in the matched case (in Sect. 6.2.2).

The simulated oscillation frequency is shown in Fig. 6.8, together with the results of Eq. (6.53a) for two values of Q. Note that a small deviation from Eq. (6.53a) is expected because Q changes with the coupling factor.

6.2.6.1 Amplitude Error

Knowing the oscillation frequency, we are now able to derive the amplitude error. Subtracting Eq. (6.22c) from Eq. (6.22d), we obtain

$$\begin{aligned} \frac{\mathrm{d}\phi_2}{\mathrm{d}t} - \frac{\mathrm{d}\phi_1}{\mathrm{d}t} &= \frac{\omega^2 - \omega_{02}^2(1-\alpha_2) - \omega^2 + \omega_{01}^2(1-\alpha_1)}{2\omega} \\ &- \frac{1}{2}\left(\alpha_{K2}\frac{I_{o1}}{I_{o2}} + \alpha_{K1}\frac{I_{o2}}{I_{o1}}\right)\sin(\Delta\phi). \end{aligned} \qquad (6.54)$$

At steady state, the derivatives are zero. If one assumes that the oscillations are nearly in quadrature ($\sin\Delta\phi \approx 1$), Eq. (6.54) is reduced to:

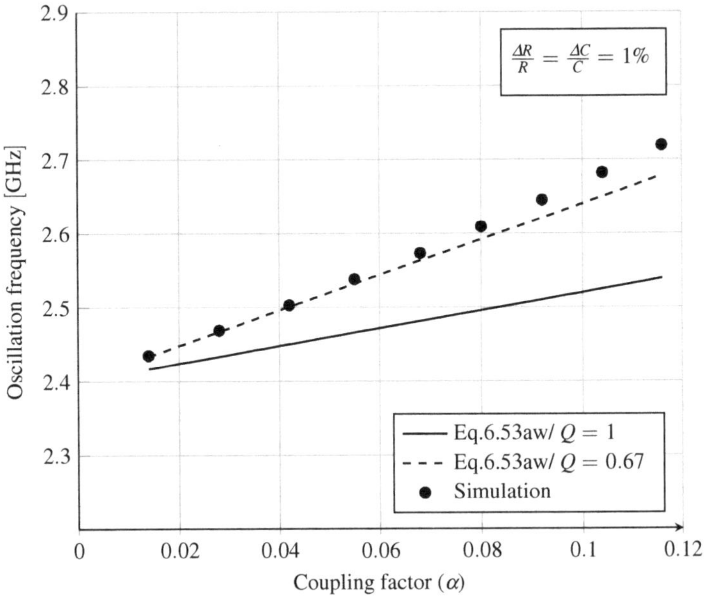

Fig. 6.8 Oscillation frequency of the capacitive coupling quadrature oscillator

$$-\Delta\omega - \frac{1}{2}\left(\alpha_{K2}\frac{I_{o1}}{I_{o2}} + \alpha_{K1}\frac{I_{o2}}{I_{o1}}\right) \approx 0. \tag{6.55}$$

where $\Delta\omega$ is the free-running frequencies' mismatch given by:

$$\Delta\omega = \frac{\omega_{02}^2(1-\alpha_2) - \omega_{01}^2(1-\alpha_1)}{2\omega}. \tag{6.56}$$

The left-hand side of Eq. (6.55) has two terms. To obtain the amplitude error, we will express each term as a function of the mismatches and the amplitude error. Using $\omega_{0i}^2(1-\alpha_i) = 1/(L_iC_i)$ and substituting this result into Eq. (6.56), the free-running frequency mismatch can be approximated by:

$$\Delta\omega = \frac{1}{2\omega}\left(\frac{1}{L_2'C_2'} - \frac{1}{L_1'C_1'}\right). \tag{6.57}$$

Expanding Eq. (6.57) results in:

$$\Delta\omega \approx -\frac{\omega_0^2(1-\alpha)}{2\omega}\left[\left(\frac{\Delta C}{C}\right)(1-\alpha) + \left(\frac{\Delta R}{R}\right)\right], \tag{6.58}$$

Equation (6.58) shows that the frequency mismatch (difference between the free-running frequencies) is dependent on the capacitances' and resistances' mismatches.

Let us now simplify the second term on the left-hand side of Eq. (6.55). For $\alpha_{K1} = -2(R_2\alpha_1 + R_1\alpha_2)/L'_{e1}$ and $\alpha_{K2} = 2(R_2\alpha_1 + R_1\alpha_2)/L'_{e2}$, assuming small mismatches and using the approximations ($\alpha_1 + \alpha_2 \approx 2\alpha$) and ($\alpha_1 - \alpha_2 \approx \frac{\Delta C}{C}(1-\alpha)\alpha$), we can obtain

$$-\frac{1}{2}\left(\alpha_{K2}\frac{I_{o1}}{I_{o2}} + \alpha_{K1}\frac{I_{o2}}{I_{o1}}\right)\sin\Delta\phi, \tag{6.59}$$

where $\alpha_{K1} = -2(R_2\alpha_1 + R_1\alpha_2)\frac{2}{L_{e1}}$ and $\alpha_{K2} = 2(R_2\alpha_1 + R_1\alpha_2)\frac{2}{L_{e2}}$. Expanding α_{K1} and α_{K2} gives

$$\alpha_{K1} = -\frac{4R}{L_{e1}}\left[(\alpha_1 + \alpha_2) + \left(\frac{\Delta R}{2R}\right)(\alpha_1 - \alpha_2)\right],$$

which can be simplified if we use the approximations $\alpha_1 + \alpha_2 \approx 2\alpha$ and $\alpha_1 - \alpha_2 \approx \frac{\Delta C}{C}(1-\alpha)\alpha$,

$$\alpha_{K1} \approx -\frac{4R}{L_{e1}}\left[2\alpha + \left(\frac{\Delta R}{2R}\right)\left(\frac{\Delta C}{C}\right)\alpha(1-\alpha)\right]. \tag{6.60}$$

For small mismatches, the second term on the right-hand side of Eq. (6.60) can be neglected, which results in:

$$\alpha_{K1} \approx -\frac{8R}{L_{e1}}\alpha. \tag{6.61}$$

Similarly,

$$\alpha_{K2} \approx \frac{8R}{L_{e2}}\alpha. \tag{6.62}$$

Substituting Eq. (6.61) and Eq. (6.62) into the second term on the left-hand side of Eq. (6.59), we obtain

$$-\frac{1}{2}\left(\alpha_{K2}\frac{I_{o1}}{I_{o2}} + \alpha_{K1}\frac{I_{o2}}{I_{o1}}\right)\sin\Delta\phi \approx -2R\alpha\left(\frac{L'_1 I_{o1}^2 - L'_2 I_{o2}^2}{L'_1 L'_2 I_{o1} I_{o2}}\right). \tag{6.63}$$

Using $I_{o1}^2 \approx I_{osc}^2(1-\epsilon_A)$, $I_{o2}^2 \approx I_{osc}^2(1+\epsilon_A)$ and $I_{o1}I_{o2} \approx I_{osc}^2\left(1 - \frac{\epsilon_A^2}{4}\right) \approx I_{osc}$, the Eq. (6.63) is reduced to:

$$-\frac{1}{2}\left(\alpha_{K2}\frac{I_{o1}}{I_{o2}}+\alpha_{K1}\frac{I_{o2}}{I_{o1}}\right)\sin\Delta\phi\approx-2R\alpha\left[\frac{L_1'-L_2'}{L_1'L_2'}-\frac{L_1'+L_2'}{L_1'L_2'}\epsilon_A\right]. \tag{6.64}$$

Substituting Eq. (6.46) in Eq. (6.64) gives

$$-\frac{1}{2}\left(\alpha_{K2}\frac{I_{o1}}{I_{o2}}+\alpha_{K1}\frac{I_{o2}}{I_{o1}}\right)\sin\Delta\phi\approx\frac{2R}{L}\alpha\left[\left(\frac{\Delta R}{R}\right)+2\epsilon_A\right]. \tag{6.65}$$

Finally, with both terms simplified, substituting Eq. (6.58) and Eq. (6.65) into Eq. (6.55) gives

$$\frac{\omega_0^2(1-\alpha)}{2\omega}\left[\left(\frac{\Delta R}{R}\right)+\left(\frac{\Delta C}{C}\right)(1-\alpha)\right]\approx-\frac{2R\alpha}{L}\left[\frac{\Delta R}{R}+2\epsilon_A\right]. \tag{6.66}$$

Using $\frac{2R}{L}=\frac{\omega_0}{Q}$ and $\omega\approx\omega_0\left(1+\frac{2-Q}{2Q}\alpha\right)$ for $\Delta\phi\approx\frac{\pi}{2}$, the above equation becomes

$$(1-\alpha)\left[\left(\frac{\Delta R}{R}\right)+\left(\frac{\Delta C}{C}\right)(1-\alpha)\right]\approx\mp\frac{2\left(1+\frac{2-Q}{2Q}\alpha\right)\alpha}{Q}\left(\frac{\Delta R}{R}\right)\mp\frac{4\left(1+\frac{2-Q}{2Q}\alpha\right)\alpha}{Q}\epsilon_A, \tag{6.67}$$

If Eq. (6.67) is solved with respect to the amplitude error, ϵ_A, the result will be

$$\epsilon_A\approx-\frac{1}{4}\left[1+\frac{(1-\alpha)}{(2+\alpha)\alpha}\right]\left(\frac{\Delta R}{R}\right)-\frac{1}{4}\left[\frac{(1-\alpha)^2}{(2+\alpha)\alpha}\right]\left(\frac{\Delta C}{C}\right). \tag{6.68}$$

Equation (6.68) shows that the amplitude error increases with both resistance and capacitance mismatches. The error is reduced by substantially increasing the coupling factor. The resistance mismatch has a slightly higher impact on the amplitude error. In addition, the resistance mismatch sets the lower limit of the amplitude error: even for very large values of the coupling factor, the amplitude error cannot be less than a quarter of the resistance mismatch.

In comparison with the simulations' results, Eq. (6.68) gives a more conservative result (approximately doubles the amplitude error), as shown in Fig. 6.9. Despite the higher values given by Eq. (6.68), the trend follows the simulation results diverging strongly only for low values of coupling factors.

6.2.6.2 Phase Error

We obtain the phase error by subtracting Eq. (6.22b) from Eq. (6.22a). This gives

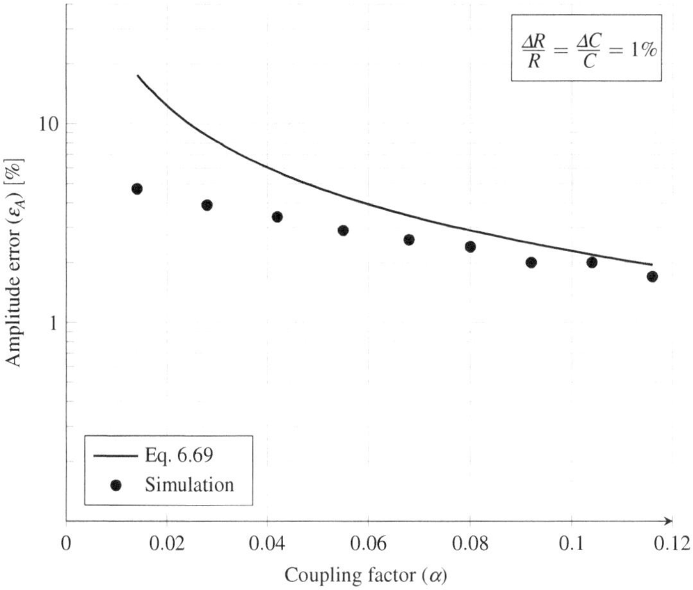

Fig. 6.9 Simulation results for the amplitude error of the capacitive coupling

$$(\gamma_0 - \delta_0) - \frac{1}{4}\left(\gamma_2 I_{o2}^2 - \delta_2 I_{o1}^2\right) + \frac{1}{2}\left(\alpha_{K2}\frac{I_{o1}}{I_{o2}} - \alpha_{K1}\frac{I_{o2}}{I_{o1}}\right)\cos(\Delta\phi) = 0. \tag{6.69}$$

Writing the first term in the left-hand side of Eq. (6.69) as a function of the resistance and capacitance mismatches, $\Delta R/R$ and $\Delta C/C$, respectively, results in:

$$(\gamma_0 - \delta_0) = \frac{R\left(1 + \frac{\Delta R}{2R}\right)\left(1 - \frac{C_d}{C_2'}\right) - g_{m0}{}^{-1}}{L\left(1 + \frac{\Delta R}{2R}\right)} - \frac{R\left(1 - \frac{\Delta R}{2R}\right)\left(1 - \frac{C_d}{C_1'}\right) - g_{m0}{}^{-1}}{L\left(1 - \frac{\Delta R}{2R}\right)} \tag{6.70}$$

Assuming small mismatches $\left(\frac{\Delta R}{2R}\right)^2 \ll 1$, and reducing both terms, at the right-hand side, to a common denominator yields

$$(\gamma_0 - \delta_0) \approx \frac{RC_d\left(\frac{1}{C_1'} - \frac{1}{C_2'}\right) + g_{m0}{}^{-1}\left(\frac{\Delta R}{R}\right)}{L}. \tag{6.71}$$

With $\frac{1}{C_1'} - \frac{1}{C_2'} \approx \frac{1}{C}\left(\frac{\Delta C}{C}\right)(1-\alpha)^2$, Eq. (6.71) is simplified to:

$$\begin{aligned}(\gamma_0 - \delta_0) &\approx \frac{RC_d\left(\frac{\Delta C}{C}\right)(1-\alpha)^2 + g_{m0}{}^{-1}C\left(\frac{\Delta R}{R}\right)}{LC} \\ &\approx \omega_0^2(1-\alpha)\left[RC_d\left(\frac{\Delta C}{C}\right)(1-\alpha) + g_{m0}{}^{-1}C\left(\frac{\Delta R}{R}\right)\right].\end{aligned} \tag{6.72}$$

For the second term on the left-hand side of Eq. (6.69), using $I_{o1} = I_{osc}\left(1 - \frac{\epsilon_A}{2}\right)$ and $I_{o2} = I_{osc}\left(1 + \frac{\epsilon_A}{2}\right)$, we obtain

$$\frac{1}{4}\left(\gamma_2 I_{o2}^2 - \delta_2 I_{o1}^2\right) = \frac{I_{osc}^2}{4}\left[\left(1 + \epsilon_A + \frac{\epsilon_A^2}{4}\right)\gamma_2 - \left(1 - \epsilon_A + \frac{\epsilon_A^2}{4}\right)\delta_2\right]$$
$$\approx \frac{I_{osc}^2}{4}\left[(\gamma_2 - \delta_2) + (\gamma_2 + \delta_2)\,\epsilon_A\right] \tag{6.73}$$

Substituting δ_2 and γ_2 by their equations as a function of the circuit elements results in:

$$\frac{1}{4}\left(\gamma_2 I_{o2}^2 - \delta_2 I_{o1}^2\right) \approx \frac{3K^2 I_{osc}^2}{4}\left[\left(\frac{R_2\left(1 - \frac{C_d}{C_2'}\right)}{L_2'} - \frac{R_1\left(1 - \frac{C_d}{C_1'}\right)}{L_1'}\right)\right.$$
$$\left. + \left(\frac{R_2\left(1 - \frac{C_d}{C_2'}\right)}{L_2'} + \frac{R_1\left(1 - \frac{C_d}{C_1'}\right)}{L_1'}\right)\epsilon_A\right] \tag{6.74}$$

Since $K = \frac{1}{4I}$, $R_1 = R\left(1 - \frac{\Delta R}{2R}\right)$, $R_2 = \left(1 + \frac{\Delta R}{2R}\right)$, $L_1' = L\left(1 - \frac{\Delta R}{2R}\right)$ and $L_2' = L\left(1 + \frac{\Delta R}{2R}\right)$, the second term of Eq. (6.69) is reduced to:

$$\frac{1}{4}\left(\gamma_2 I_{o2}^2 - \delta_2 I_{o1}^2\right) \approx \frac{3I_{osc}^2 R}{4^3 I^2 LC}\left[C_d\left(\frac{\Delta C}{C}\right)(1-\alpha)^2 + 2\left(C - C_d\left(1-\alpha\right)\right)\epsilon_A\right] \tag{6.75}$$

Note that $I_{osc} \approx 8I\sqrt{\frac{Rg_{m0} - \frac{C}{C - C_d}}{3Rg_{m0}}}$, which substituted into Eq. (6.75), yields

$$\frac{1}{4}\left(\gamma_2 I_{o2}^2 - \delta_2 I_{o1}^2\right)$$
$$\approx \frac{Rg_{m0} - \frac{C}{C - C_d}}{Rg_{m0}}\omega_0^2 R\left[C_d\left(\frac{\Delta C}{C}\right)(1-\alpha)^2 + 2\left(C - C_d\left(1-\alpha\right)\right)\epsilon_A\right] \tag{6.76}$$

A similar procedure should be applied to the third term on the left-hand side of Eq. (6.69). Substituting α_{K1} and α_{K2} by Eq. (6.61) and Eq. (6.62) into Eq. (6.76) results in:

$$\frac{1}{2}\left(\alpha_{K2}\frac{I_{o1}}{I_{o2}} - \alpha_{K1}\frac{I_{o2}}{I_{o1}}\right)\cos(\Delta\phi) = (R_2\alpha_1 + R_1\alpha_2)\left(\frac{1}{L_2'}\frac{I_{o1}}{I_{o2}} + \frac{1}{L_1'}\frac{I_{o2}}{I_{o1}}\right)$$
$$\cos(\Delta\phi) = 0 \tag{6.77}$$

Multiplying both sides of Eq. (6.77) by I_{o1} and I_{o2} results in $I_{o1}I_{o2}$ as common denominator:

$$\frac{1}{2}\left(\alpha_{K2}\frac{I_{o1}}{I_{o2}} - \alpha_{K1}\frac{I_{o2}}{I_{o1}}\right)\cos(\Delta\phi)$$
$$= (R_2\alpha_1 + R_1\alpha_2)\left(\frac{1}{L_2'}\frac{I_{o1}^2}{I_{o2}I_{o1}} + \frac{1}{L_1'}\frac{I_{o2}^2}{I_{o1}I_{o2}}\right)\cos(\Delta\phi) = 0 \tag{6.78}$$

Now, using Eq. (6.44) and Eq. (6.46) in Eq. (6.78), we obtain

$$\frac{1}{2}\left(\alpha_{K2}\frac{I_{o1}}{I_{o2}} - \alpha_{K1}\frac{I_{o2}}{I_{o1}}\right)\cos(\Delta\phi)$$
$$\approx \frac{R}{L}\left[(\alpha_1+\alpha_2) + \frac{\Delta R}{2R}(\alpha_1-\alpha_2)\right]\left(\frac{\left(I_{o1}^2 + I_{o2}^2\right) + \frac{\Delta R}{2R}\left(I_{o2}^2 - I_{o1}^2\right)}{I_{o2}I_{o1}}\right)\cos(\Delta\phi) \tag{6.79}$$

Furthermore, using $I_{o1}^2 \approx I_{osc}^2(1-\epsilon_A)$, $I_{o2}^2 \approx I_{osc}^2(1+\epsilon_A)$ and $I_{o1}I_{o2} \approx I_{osc}^2$, after some tedious calculations, one simplifies the third term of Eq. (6.79) to:

$$\frac{1}{2}\left(\alpha_{K2}\frac{I_{o1}}{I_{o2}} - \alpha_{K1}\frac{I_{o2}}{I_{o1}}\right)\cos(\Delta\phi)$$
$$\approx \frac{R}{L}\left[(\alpha_1+\alpha_2) + \frac{\Delta R}{2R}(\alpha_1-\alpha_2)\right]\left(2 + \frac{\Delta R}{2R}\epsilon_A\right)\cos(\Delta\phi) \tag{6.80}$$

Note that if we assume small mismatches, the terms: $\frac{\Delta R}{2R}\epsilon_A \ll 2$, $\frac{\Delta R}{2R}\frac{\Delta C}{2C}(1-\alpha) \ll 2$, $\alpha_1+\alpha_2 \approx 2\alpha$ and $\alpha_1-\alpha_2 \approx \alpha(1-\alpha)\left(\frac{\Delta C}{C}\right)$. Thus, Eq. (6.80) is reduced to:

$$\frac{1}{2}\left(\alpha_{K2}\frac{I_{o1}}{I_{o2}} - \alpha_{K1}\frac{I_{o2}}{I_{o1}}\right)\cos(\Delta\phi) \approx 4RC\alpha\omega_0^2\cos(\Delta\phi) \tag{6.81}$$

Finally, substituting Eq. (6.71), Eq. (6.76) and Eq. (6.81) into Eq. (6.69) and solving the result with respect to $\cos\Delta\phi$, we obtain

$$4RC\alpha\cos(\Delta\phi) = (1-\alpha)\left[RC_d\left(\frac{\Delta C}{C}\right)(1-\alpha) + g_{m0}{}^{-1}C\left(\frac{\Delta R}{R}\right)\right] + \\ -\frac{Rg_{m0} - \frac{C}{C-C_d}}{Rg_{m0}}\left[RC_d\left(\frac{\Delta C}{C}\right)(1-\alpha)^2 + 2R\left(C - C_d(1-\alpha)\right)\epsilon_A\right] \tag{6.82}$$

Rearranging the terms in Eq. (6.82) results in:

$$\cos(\Delta\phi) = \frac{(1-\alpha)^2}{\alpha}\frac{1}{4Rg_{m0}}\left[\frac{C_d}{(C-C_d)}\left(\frac{\Delta C}{C}\right) + \frac{1}{(1-\alpha)}\left(\frac{\Delta R}{R}\right) \\ -2\left(Rg_{m0} - \frac{C}{C-C_d}\right)\left(\frac{1}{(1-\alpha)^2} - \frac{C_d}{C}\frac{1}{(1-\alpha)}\right)\epsilon_A\right] \tag{6.83}$$

The phase difference is

$$\Delta\phi = \frac{\pi}{2} + \epsilon_\phi, \tag{6.84}$$

where ϵ_ϕ is the phase error.

The cosine of the phase difference is equal to the sine of the phase error, i.e. $\cos\Delta\phi = \sin\epsilon_\phi$. Moreover, assuming that the phase error is small, $\cos\Delta\phi$ is approximately equal to the phase error:

$$\epsilon_\phi \approx \frac{(1-\alpha)^2}{\alpha}\frac{1}{4Rg_{m0}}\left[\frac{C_d}{C-C_d}\left(\frac{\Delta C}{C}\right) + \frac{1}{(1-\alpha)}\left(\frac{\Delta R}{R}\right) \\ -2\frac{Rg_{m0} - \frac{C}{C-C_d}}{(1-\alpha)^2}\left(1 - \frac{C_d}{C}(1-\alpha)\right)\epsilon_A\right] \tag{6.85}$$

For the specific case of $C = 2C_d$, the phase error is reduced to:

$$\epsilon_\phi \approx \frac{(1-\alpha)^2}{\alpha}\frac{1}{4Rg_{m0}}\left[\left(\frac{\Delta C}{C}\right) + \frac{1}{(1-\alpha)}\left(\frac{\Delta R}{R}\right) - \frac{Rg_{m0} - 2}{(1-\alpha)^2}(1+\alpha)\epsilon_A\right]. \tag{6.86}$$

Note that Eq. (6.86) is in radians. To obtain the phase error in degrees, we multiply Eq. (6.86) by $\frac{180}{\pi}$:

$$\epsilon_{\phi(\text{deg})} \approx \frac{(1-\alpha)^2}{\alpha}\frac{45}{\pi Rg_{m0}}\left[\left(\frac{\Delta C}{C}\right) + \frac{1}{(1-\alpha)}\left(\frac{\Delta R}{R}\right) - \frac{Rg_{m0} - 2}{(1-\alpha)^2}(1+\alpha)\epsilon_A\right]. \tag{6.87}$$

Equation (6.87) shows that the phase error increases with both resistance and capacitance mismatches and decreases with the amplitude error. The error diminishes substantially with the increase of the coupling factor. The simulations'

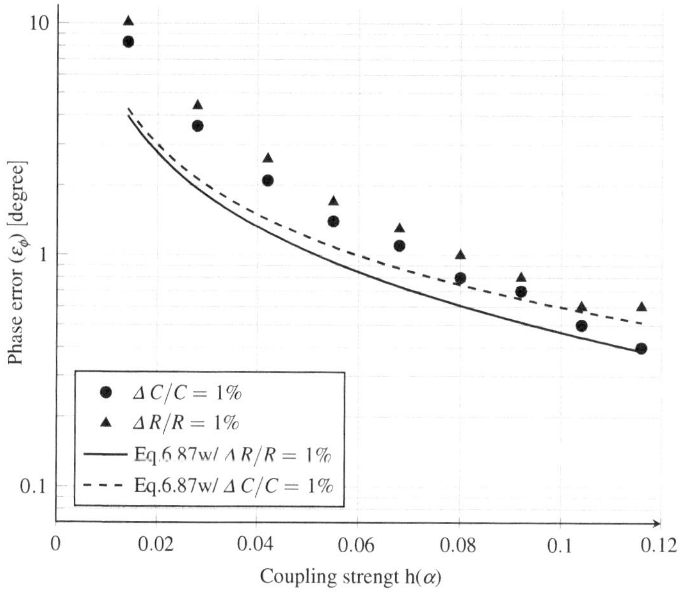

Fig. 6.10 Simulated phase error

results are in agreement with the theory equation (6.87), as shown in Fig. 6.10. The small difference between the simulation and theoretical results is explained by the approximations used.

Contrary to the *LC*-oscillator [11], the cross-coupled *RC*-oscillator has a low sensitivity of the phase noise to the coupling factor. This means that, although the phase error reduces substantially with the increase of the coupling factor, the phase noise has a negligible variation (about 1 dB), as shown in Fig. 6.11. Thus, the designer can focus on the reduction of the phase error because the penalty in the phase noise is negligible.

6.3 Experimental Results

To validate the theory, a 2.4-GHz capacitive coupled QVCO with variable coupling capacitances was fabricated in UMC 0.13 μm CMOS process. The circuit schematic is shown in Fig. 6.12. The coupling capacitances are 3-bit binary weighted capacitor arrays, as shown in Fig. 6.13a. Each capacitor array has a step of 20 fF with 3 bits allowing a capacitance variation range from approximately 0 fF (not coupled) up to 140 fF coupling. The prototype die microphotograph is shown in Fig. 6.13d. A second prototype was made with a single capacitance value to minimize the area. The die microphotograph of the second prototype is shown in Fig. 6.13c. This prototype has a switch to turn the coupling on and off. Each prototype die

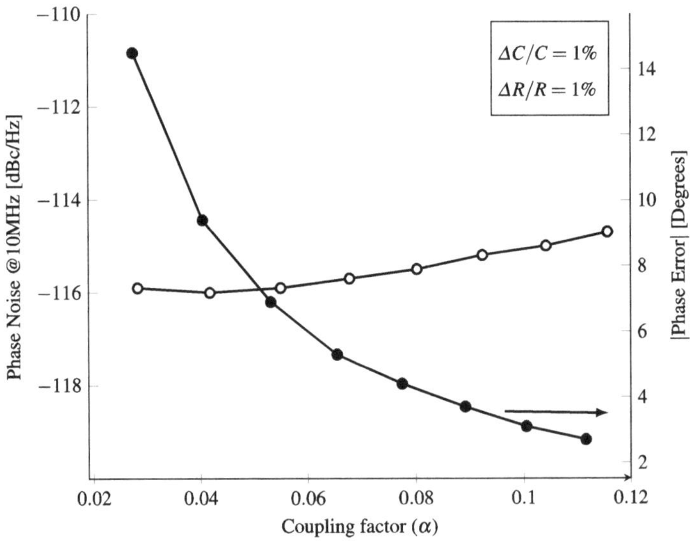

Fig. 6.11 Phase noise and phase error

was bondwired to a printed circuit board (PCB) making the RF signals accessible through four SMA connectors, as shown in the photograph of Fig. 6.13b. We refer to this PCB as the daughterboard, since a second PCB (the motherboard) is required to provide the power supplies and control signals.

The dimensions of the oscillators core transistors (M_1, M_2, M_3, and M_4) are $W = 7.2\,\mu\text{m}$, and $L = 120\,\text{nm}$. The dimensions of the current source transistors (M_9, M_{11}, M_{12}, and M_{14}) are $W = 7.2\,\mu\text{m}$, and $L = 360\,\text{nm}$. The resistors were implemented with PMOS transistors, operating in the triode region, with $W = 5.4\,\mu\text{m}$, and $L = 120\,\text{nm}$. The timing capacitors are of MiM type, with an area of $20\,\mu\text{m} \times 20\,\mu\text{m}$, resulting in the capacitance of 431.7 fF. The supply voltage is 1.2 V, and the bias current is 1.8 mA, which results in 8.64 mW power dissipation. The layout of the circuit occupies an area of $430\,\mu\text{m} \times 180\,\mu\text{m}$ (without pads).

Figure 6.14 shows the measured oscillation frequencies when the oscillators are free running, i.e. $C_X \approx 0\,\text{fF}$, represented in the figure by the triangles, and coupled with $C_X = 20\,\text{fF}$ (dots). The gap between the two results clearly indicates that the oscillation frequency increases when the oscillators are coupled, which is consistent with the theory.

The relation between the oscillation frequency and the coupling strength is shown in Fig. 6.15, where the dots are the measurement results and the 3-digit code, beside each dot, are the corresponding logic states of the switches (S_2, S_1, S_0). As expected, the oscillation frequency increases almost linearly with the coupling capacitance C_X and the amplitude of the output voltage decreases. However, note that the frequency increase is higher than expected due to the parasitic capacitances and low quality factor (below 1). Extracting the coupling capacitance value from

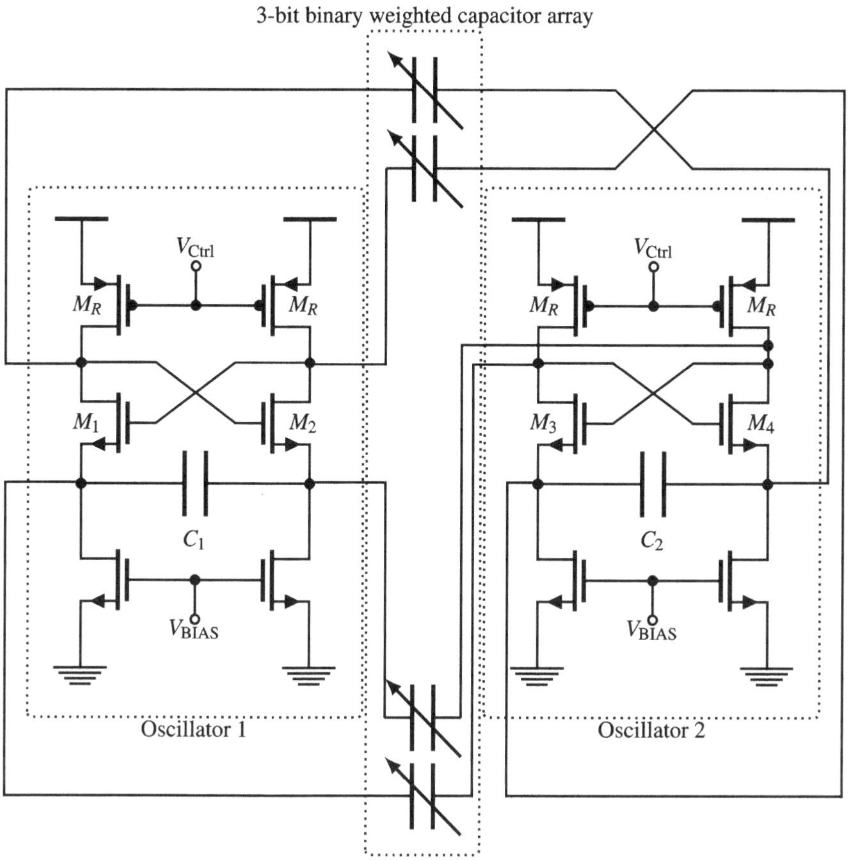

Fig. 6.12 Prototype circuit of the capacitive coupling oscillator

the trend line (solid line) yields $C_X \approx 92$ fF. This indicates that the parasitics have a strong influence on the coupling capacitances.

The measured phase noise is −115.1 dBc/Hz @ 10 MHz, as shown in Fig. 6.16. To guarantee a nearly sinusoidal output, all the measurements were made with the power of the third harmonic 25 dB below that of the fundamental.

To compare this oscillator with others, with similar topology, we use the conventional FoM [12]:

$$\text{FoM} = \text{PN} + 10\log\left(\left(\frac{\Delta f}{f}\right)^2 \frac{P_{\text{DC}}}{P_{\text{ref}}}\right), \tag{6.88}$$

where PN is the phase noise, P_{DC} is the dissipated power in (mW), P_{ref} is the reference power (typically 1 mW), f is the oscillation frequency and Δf is the frequency offset.

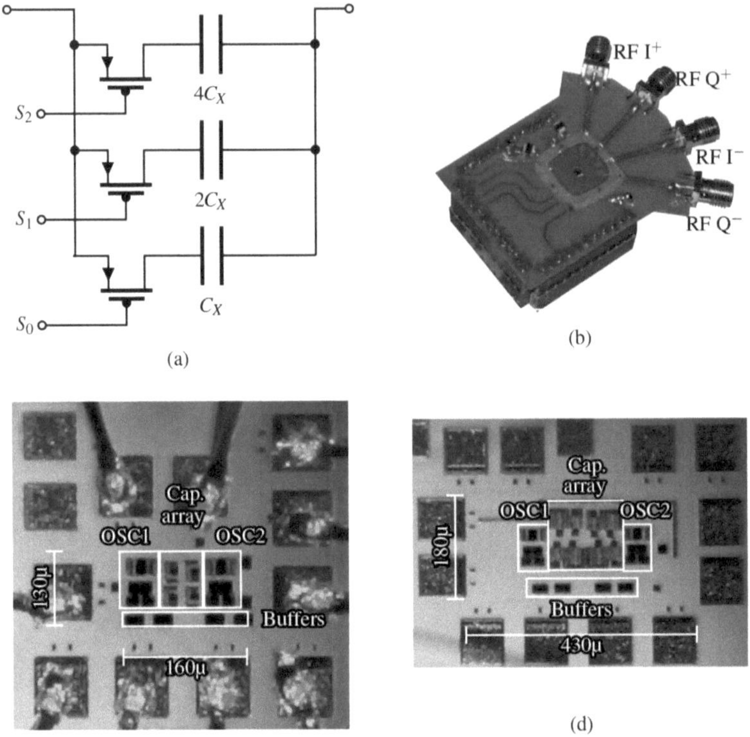

Fig. 6.13 3-bit binary weighted capacitor array (**a**), photo of the daughterboard (**b**), the microphotos of the capacitive coupling QVCOs without capacitor array (**c**) and with capacitor array (**d**)

Table 6.1 gives a comparison among the state-of-the-art quadrature RC-oscillators. A figure-of-merit (FoM) of -154.8 dBc/Hz is obtained for the oscillator described here with a power of 8.64 mW, which is the best performance for a QVCO with nearly sinusoidal output.

6.4 Conclusion

The capacitive cross-coupling of RC-oscillators was analysed theoretically. Simulation and measurement results confirm that this coupling scheme is a viable solution to generate quadrature outputs. In comparison with the active coupling schemes, it reduces the noise and power dissipation.

Several simulations using real MOS transistor models have been performed to validate the theory. Simulations, using SpectreRF, confirmed the inverse proportionality of the phase error and amplitude mismatch to the coupling factor. A phase error

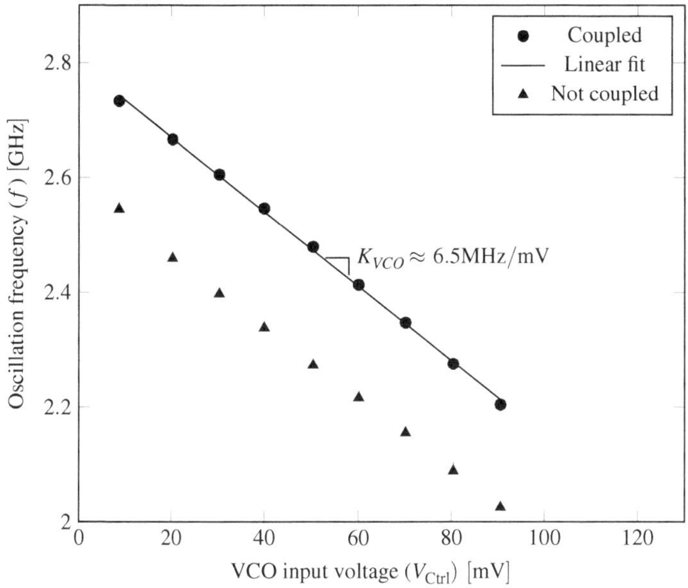

Fig. 6.14 Frequency of oscillation with the oscillators uncoupled and coupled ($C_X = 20\,\text{fF}$)

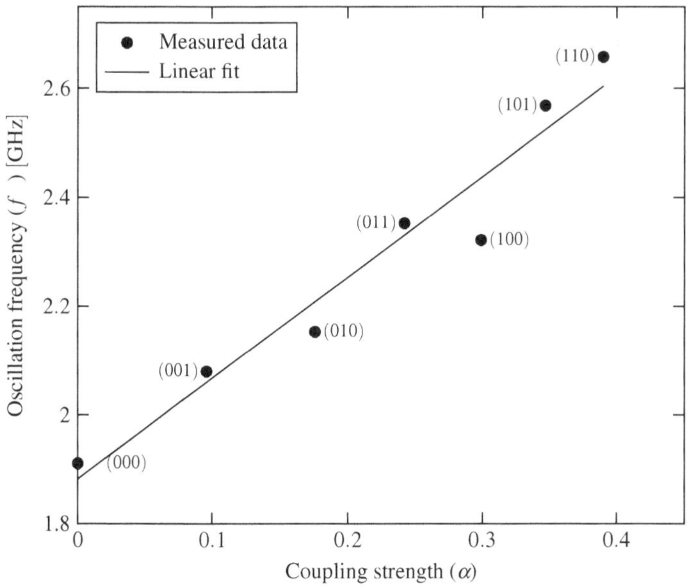

Fig. 6.15 Relation between the oscillation frequency and the coupling strength

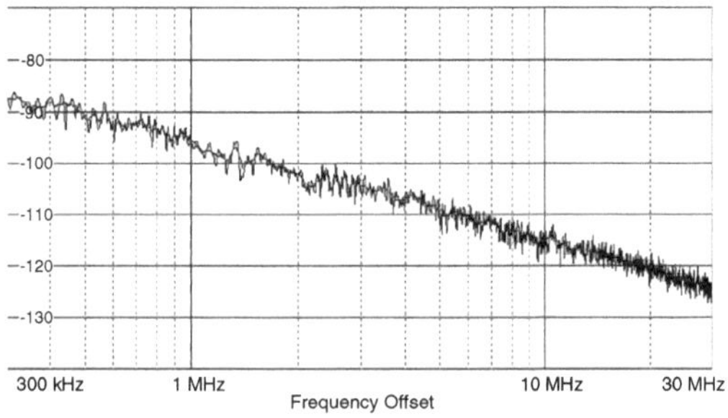

Fig. 6.16 Measured phase noise

Table 6.1 Comparison of state-of-the-art nearly sinusoidal RC-oscillators with the same circuit topology

Reference	Frequency (MHz)	PN @ Δf (dBc/Hz)	Power dissipation (mW)	FoM (dBc/Hz)	IQ	Area (μm × μm)
[13]	920	−102@1 MHz	9.9	-151.3	No	N/A
[14]	5000	−97.1@1 MHz	54	-153.8	Yes	350 × 700
[15]	2290	−105@1 MHz	54.72	-154.8	Yes	300 × 350
This work	2850	−115.1@10 MHz	8.64	-154.8	Yes	160 × 130

below 1% and an amplitude mismatch lower than 1% are obtained with a coupling capacitance of about 20% of the oscillator's capacitance value.

Simulations also showed that, contrarily to the LC-oscillator, the cross-coupled RC-oscillator has low sensitivity of the phase noise to the coupling factor. This means that, although the phase error reduces substantially with the increase of the coupling factor, the phase noise has a negligible variation (about 1 dB). Thus, the designer can concentrate on the reduction of the phase error, because the penalty in the phase noise is negligible.

A circuit prototype was designed, which has a phase noise of −115.1 dBc/Hz @10 MHz (this is about 3 dB improvement in comparison with a single RC-oscillator). The increase of power is only marginal, leading to a FoM of −154.8 dBc/Hz. These results are consistent with the noiseless feature of the capacitive coupling and are comparable to those of the best state-of-the-art RC-oscillators in the GHz range, but with the lowest power consumption (about 9 mW).

The proportionality between the oscillation frequency and the coupling factor was confirmed in a prototype with a variable capacitor array used in the coupling of two RC-oscillators. Although, in practice, an increase of the oscillation frequency was observed. In theory, there are two operation modes, in one of which the frequency decreases with the increase of the coupling factor, but this mode was not observed because it was not possible in the prototype to select which oscillator starts first.

Finally, the theory presented led to the interesting result that the amplitude mismatch is related to the phase error. This relation indicates that an automatic phase-error minimization circuit can be implemented, consisting of a feedback loop that measures the amplitude mismatch (using two peak detectors) and adjusts the oscillators' current sources until the amplitudes are matched. This will ensure a reduction of the phase error.

References

1. A. Willson, Energy circulation quadrature LC-VCO, in *IEEE International Symposium on Circuits and Systems* (IEEE, Kos, 2006), p. 4
2. S. Gierkink, S. Levantino, R. Frye, C. Samori, V. Boccuzzi, A low-phase-noise 5-GHz CMOS quadrature VCO using superharmonic coupling. IEEE J. Solid State Circuits **38**(7), 1148–1154 (2003)
3. W. Li, K.-K.M. Cheng, A CMOS transformer-based current reused SSBM and QVCO for UWB application. IEEE Trans. Microw. Theory Tech. **61**(6), 2395–2401 (2013)
4. C.T. Fu, H.C. Luong, A 0.8-V CMOS quadrature LC VCO using capacitive coupling, in *IEEE Asia Solid-State Circuits Conference (ASSCC'07)* (2007), pp. 436–439
5. L.B. Oliveira, I.M. Filanovsky, A. Allam, J.R. Fernandes, Synchronization of two LC-oscillators using capacitive coupling, in *IEEE International Symposium on Circuits and Systems (ISCAS'08)* (2008), pp. 2322–2325
6. J. Casaleiro, L.B. Oliveira, I. Filanovsky, Low-power and low-area CMOS quadrature RC oscillator with capacitive coupling, in *IEEE International Symposium on Circuits and Systems (ISCAS'12)* (2012), pp. 1488–1491
7. H. Tong, S. Cheng, Y.-C. Lo, A.I. Karsilayan, J. Silva-Martinez, An LC quadrature VCO using capacitive source degeneration coupling to eliminate bi-modal oscillation. IEEE Trans. Circuits Syst. I **59**(9), 1871–1879 (2012)
8. J. Casaleiro, L.B. Oliveira, I.M. Filanovsky, A quadrature RC-oscillator with capacitive coupling. Integr. VLSI J. **52**, 260–271 (2016)
9. P.R. Gray, P.J. Hurst, S.H. Lewis, R.G. Meyer, *Analysis and Design of Analog Integrated Circuits*, 5th edn. (Wiley, New York, 2009)
10. D.W. Jordan, P. Smith, *Nonlinear Ordinary Differential Equations: An Introduction for Scientists and Engineers*, 4th edn. (Oxford University Press, New York, 2007)
11. L. Romano, S. Levantino, C. Samori, A.L. Lacaita, Multiphase LC oscillators. IEEE Trans. Circuits Syst. I **53**(7), 1579–1588 (2006)
12. J.O. Plouchart, H. Ainspan, M. Soyuer, A. Ruehli, A fully-monolithic SiGe differential voltage-controlled oscillator for 5 GHz wireless applications, in *IEEE Radio Frequency Integrated Circuits (RFIC) Symp. - Dig. Papers* (2000), pp. 57–60
13. B. Razavi, A study of phase noise in CMOS oscillators. IEEE J. Solid State Circuits **31**(3), 331–343 (1996)
14. L.B. Oliveira, A. Allam, I.M. Filanovsky, J.R. Fernandes, C.J.M. Verhoeven, M.M. Silva, Experimental comparison of phase-noise in cross-coupled RC- and LC-oscillators. Int. J. Circuit Theory Appl. **38**, 681–688 (2010)
15. L.B. Oliveira, J.R. Fernandes, M.M. Silva, I.M. Filanovsky, C.J.M. Verhoeven, Experimental evaluation of phase-noise and quadrature error in a CMOS 2.4 GHz relaxation oscillator, in *IEEE International Symposium Circuits and Systems (ISCAS'07)* (2007), pp. 1461–1464

Chapter 7
Two-Integrator Oscillator

7.1 Introduction

In the previous chapters, two coupled RC-oscillators were analysed, the first with active coupling using transconductance amplifiers, in Chap. 5, and the second with passive coupling using capacitors, in Chap. 6. The results show that these coupled oscillators are a viable solution to generate quadrature outputs with a phase error below 1°. However, the poor phase noise and possibility of bimodal oscillations are the serious disadvantages of these oscillators. The oscillator presented in this chapter minimizes these deficiencies, while maintaining the advantages of the coupled oscillators.

In this chapter, we analyse the two-integrator oscillator, which has a working principle fundamentally different from that of coupled oscillators. Although being an RC-oscillator (inductorless), it is a single-loop oscillator with inherent quadrature outputs. In comparison with LC-oscillators, the phase noise of a two-integrator oscillator is worse. Yet, it has better noise performance than differential cross-coupled RC-oscillators [1]. This oscillator has a wide tuning range. This is important, because a wide tuning range quasi-sinusoidal quadrature voltage-controlled oscillator (QVCO) is a key block in fully integrated multiband and multistandard RF CMOS receivers. In comparison with using many narrowband receivers, a wideband receiver has reduced cost and increased flexibility. With the widespread use of radio for internet-of-things (IoT), the demand for receivers and, therefore, QVCOs operating in ISM radio bands, has been growing. To cope with this demand, a considerable research effort has been made towards the design of suitable QVCOs. Typically, a quadrature error below 1° and a tuning range of one decade are required. Generating such accurate quadrature signals for this wide range of frequencies is challenging [1–3].

The two-integrator oscillator consists of a cascade of two integrators with a signal inversion in a feedback structure. The ideal integrators produce 90° phase shift each, thus generating the quadrature signals. However, real integrators have phase

J. C. F. d. A. Casaleiro et al., *Quadrature RC–Oscillators*, Analog Circuits and Signal Processing, https://doi.org/10.1007/978-3-030-00740-9_7

and amplitude errors. The oscillator can work in either quasi-linear (outputs nearly sinusoidal) or strongly nonlinear (triangular waveform) regimes [1, 4]. Here, we focus the study on the quasi-linear regime. The main objective of the presentation is to determine the impact of the components' mismatches on the frequency, and on the amplitude and phase errors. In [5], these errors were investigated without relating the results to the components' mismatches. Here, we investigate this relationship by expanding the approach first presented in [6].

The chapter is organized as follows. In Sect. 7.1, the oscillator model and a typical implementation of a two-integrator oscillator are presented. The derivation of the incremental model follows. The model is derived using the VDPO approximation. The two-integrator oscillator is reduced to two coupled parallel VDPOs. From the incremental model, we derive the equations of the frequency, amplitude and phase, using the negative-resistance model for the analysis. A brief explanation of the approach can be found in [1]. In Sect. 7.2, the relation between the van der Pol parameters and the circuit implementation is derived. Section 7.3 presents the simulation results. Section 7.4 provides conclusions and the discussion of possible performance improvement.

7.2 Quadrature Oscillator

The block diagram of the two-integrator oscillator is shown in Fig. 7.1. It consists of two cascaded integrators with a signal inversion in the feedback path. Each integrator is implemented by a transconductance amplifier, G_{mi}, and a capacitance, C_i, $(i = 1, 2)$. The resistance R_i represents the losses which are compensated by the

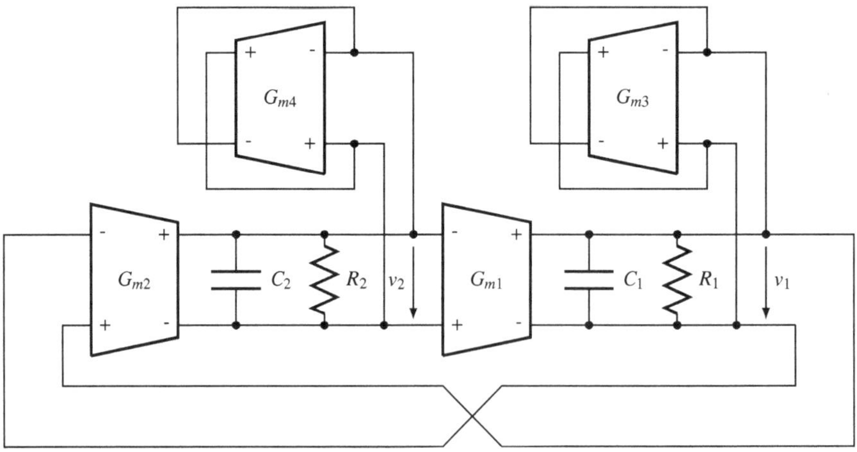

Fig. 7.1 Model of the two-integrator oscillator

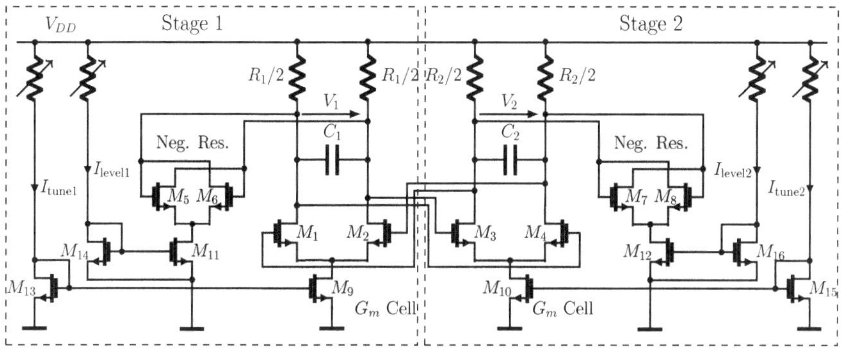

Fig. 7.2 Two-integrator oscillator

negative-resistance circuits, implemented by the transconductance amplifiers, G_{m3} and G_{m4}.

The circuit implementation of the two-integrator oscillator is shown in Fig. 7.2. The loop transconductance amplifiers are implemented by the source-coupled differential pairs, $M_{1,2}$ and $M_{3,4}$. The negative-resistance circuits are implemented by cross-coupled differential pairs, $M_{5,6}$ and $M_{7,8}$, connected in parallel with the capacitances C_1 and C_2, respectively. The resistances $R_1/2$ and the current mirror implemented by M_9 and M_{13} set the bias point of the first stage, and $R_2/2$, M_{10} and M_{15} set the bias point of the second stage.

A complete analysis of the circuit requires the inclusion of nonlinearities, which provides the amplitude limitation. As in the previous chapters, this oscillator will be also represented by the VDPO equivalent model [7, 8]. The VDPO stability was extensively studied, which also justifies approximating the two-integrator oscillator by a VDPO.

We consider that the transistors are the only elements in the circuit (Fig. 7.2) that have nonlinearities. We use for this purpose a Taylor expansion of the drain current equation to obtain the signal dependent transconductances.

7.2.1 Transconductance Amplifier

A differential transconductance amplifier is implemented by a differential pair, as shown in Fig. 7.3a. We assume that both transistors are in strong inversion and that the tail current source is ideal.

For an ideal current source, only the differential-mode analysis is relevant. The incremental model of the differential pair is shown in Fig. 7.3b. Here, G_1 and G_2 are signal-dependent transconductances (see Chap. 3), that, for convenience, we refer to, from now on, as large-signal transconductances, of M_1 and M_2, respectively. Assuming, at this stage, that there are no mismatches and that the

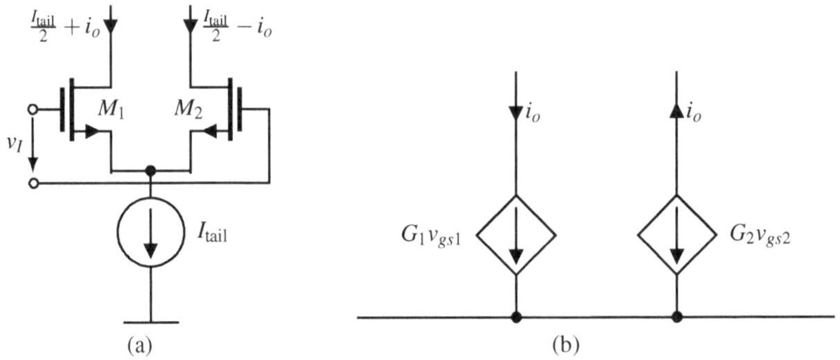

Fig. 7.3 Fully differential transconductance amplifier circuit (**a**) and incremental model (**b**)

signal is antisymmetric, i.e. $v_{gs1} = -v_{gs2} = v_i/2$, we have antisymmetric signal-dependent transconductances:

$$\begin{cases} G_1 = g_{m0} + K\dfrac{v_i}{2} & (7.1a) \\ G_2 = g_{m0} - K\dfrac{v_i}{2} & (7.1b) \end{cases}$$

Applying the KCL to the circuit, we obtain

$$\begin{cases} i_o = G_1 v_{gs1} & (7.2a) \\ i_o = -G_2 v_{gs2} & (7.2b) \\ v_i = v_{gs1} - v_{gs2} \quad . & (7.2c) \end{cases}$$

Substituting Eq. (7.2a) and Eq. (7.2b) into Eq. (7.2c), using the large-signal transconductances, given by Eq. (7.1a) and Eq. (7.1b), and solving with respect to the output current, i_o, we obtain the output current as a function of the input voltage:

$$i_o = \frac{G_1 G_2}{G_2 + G_1} v_i = \frac{g_{m0}}{2} v_i - \frac{K^2}{8 g_{m0}} v_i^3 . \tag{7.3}$$

The second term on the right-hand side of Eq. (7.3) indicates a significant distortion for high-amplitude input signals. However, for small amplitudes this term can be neglected, resulting in:

$$i_o \approx \frac{g_{m0}}{2} v_i . \tag{7.4}$$

Thus, for small amplitude the response is almost linear, as shown in Fig. 7.4. The figure shows a comparison between the theory and simulation using transistors' models of a standard CMOS technology. Note that if $i_o = I_{tail}/2$, then transistor M_1

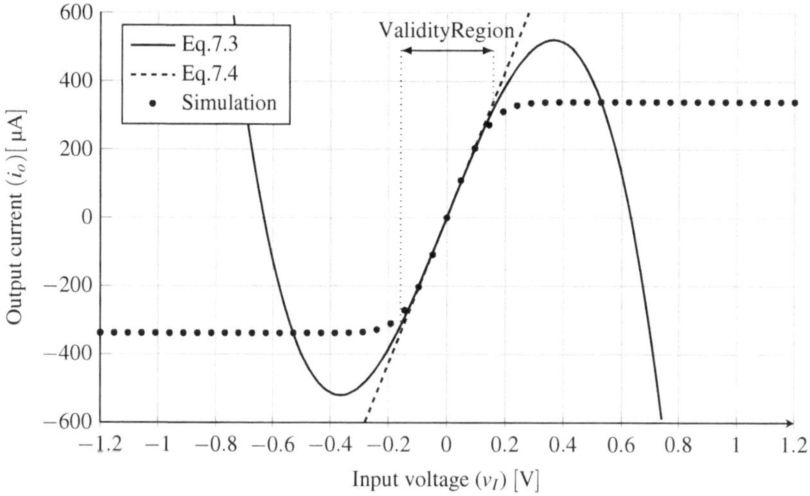

Fig. 7.4 Output current of the transconductance amplifier as a function of the differential input voltage. Transistor dimensions are $W = 14.4\,\mu\text{m}$, $L = 120\,\text{nm}$, $I_{\text{tail}} = 676\,\mu\text{A}$ and $g_{m0} = 4.28\,\text{mS}$

Fig. 7.5 Negative-resistance circuit

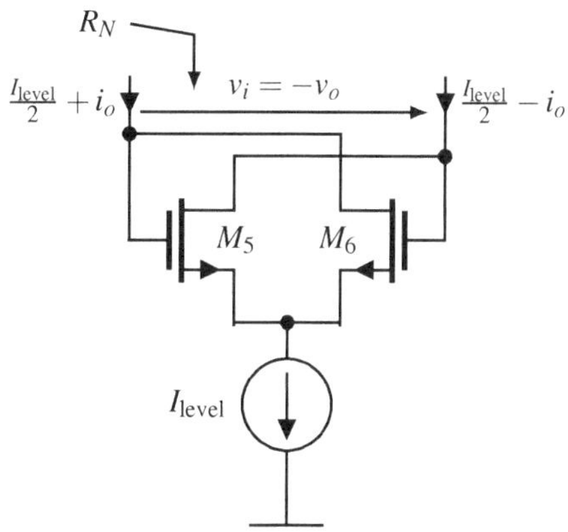

is in strong inversion but M_2 is in cutoff. Inversely, if $i_o = -I_{\text{tail}}/2$, then transistor M_2 is in strong inversion but M_1 is in cutoff. Equations (7.3) and (7.4) are valid for $|v_i| < I_{\text{tail}}/g_{m0}$. This validity region is indicated at the top of Fig. 7.4.

7.2.2 Negative-Resistance Circuit

A transconductance amplifier with the output cross-connected to the input (Fig. 7.5) behaves as a negative resistance. As in Sect. 7.2.1, only the differential-mode analysis is relevant, resulting in the incremental model shown in Fig. 7.6.

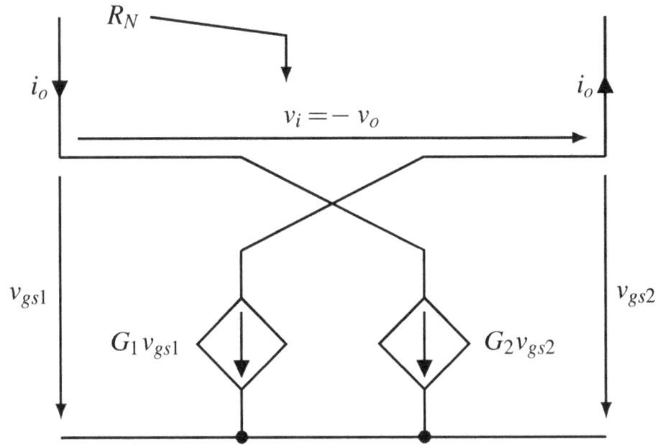

Fig. 7.6 Negative-resistance equivalent circuit

The equivalent resistance looking into the drains of M_5 and M_6 is given by the ratio between the output voltage, v_o, and the output current, i. The output current is the negative of that given by Eq. (7.3); therefore, the resistance $R_{\rm N}$ is given by:

$$R_N = \frac{v_o}{i_o} = -\frac{1}{\frac{g_{m0}}{2} - \frac{K^2}{8g_{m0}} v_o^2}. \tag{7.5}$$

Equation (7.5) shows that the circuit of Fig. 7.5 is equivalent to a negative resistance in parallel with a positive nonlinear resistance. The latter, as we will see, is responsible for the amplitude limitation.

7.2.3 Incremental Model

If the transconductance amplifiers, G_{m1} and G_{m2}, operate in the linear region, the two-integrator oscillator works linearly. However, the transconductance amplifiers of the negative-resistance circuits, G_{m3} and G_{m4}, should work in a nonlinear region to limit the amplitude. Hence, for large-signal operation each stage of the circuit in Fig. 7.2 should be modelled by a parallel RC-circuit in parallel with a nonlinear resistance, to limit the amplitude, a negative resistance to compensate the losses, and a dependent current source to represent the transconductance amplifier. The overall oscillator circuit is modeled by two coupled parallel RC-circuits, as shown in Fig. 7.7. The signal inversion is indicated by a negative transconductance in the second stage.

By applying KCL to the circuit of Fig. 7.7, we obtain

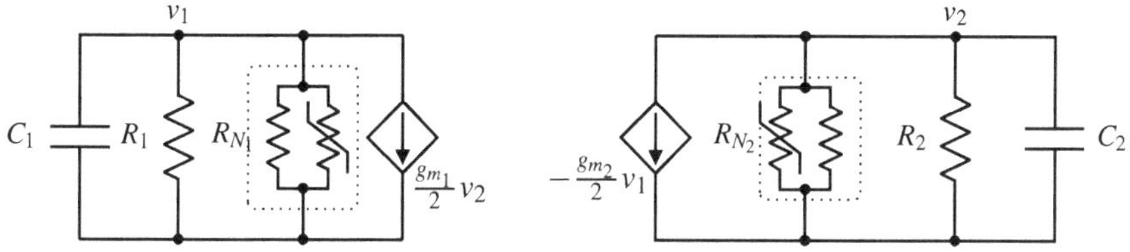

Fig. 7.7 Two-integrator small-signal equivalent circuit

$$\begin{cases} C_1\dfrac{\mathrm{d}v_1}{\mathrm{d}t} + \dfrac{1}{R_1}v_1 + \dfrac{1}{R_{N1}}v_1 + \dfrac{g_{m1}}{2}v_2 = 0 & (7.6a)\\ C_2\dfrac{\mathrm{d}v_2}{\mathrm{d}t} + \dfrac{1}{R_2}v_2 + \dfrac{1}{R_{N2}}v_2 - \dfrac{g_{m2}}{2}v_1 = 0. & (7.6b)\end{cases}$$

Dividing these equations by capacitances, C_1 and C_2, and differentiating both sides, we obtain

$$\begin{cases} \dfrac{\mathrm{d}^2v_1}{\mathrm{d}t^2} + \dfrac{1}{C_1}\left[\left(R_1^{-1} - \dfrac{g_{m0}}{2}\right) + \dfrac{3K^2}{8g_{m0}}v_1^2\right]\dfrac{\mathrm{d}v_1}{\mathrm{d}t} + \dfrac{g_{m1}}{2C_1}\dfrac{\mathrm{d}v_2}{\mathrm{d}t} = 0, & (7.7a)\\ \dfrac{\mathrm{d}^2v_2}{\mathrm{d}t^2} + \dfrac{1}{C_2}\left[\left(R_2^{-1} - \dfrac{g_{m0}}{2}\right) + \dfrac{3K^2}{8g_{m0}}v_2^2\right]\dfrac{\mathrm{d}v_2}{\mathrm{d}t} - \dfrac{g_{m2}}{2C_2}\dfrac{\mathrm{d}v_1}{\mathrm{d}t} = 0. & (7.7b)\end{cases}$$

It can be noted that the damping terms in Eq. (7.7a) and Eq. (7.7b) are similar to the damping term of the VDPO. Although each stage cannot oscillate by itself, Eqs. (7.7a) and (7.7b) are a system of van der Pol equations, and the resulting oscillator is able to oscillate.

7.2.4 *Oscillator Without Mismatches*

Now, we derive the equations of frequency, amplitude, and phase, for the steady state, assuming no mismatches between the stages. To understand how the circuit reaches the steady state, we do a stability analysis by deriving the phase-space paths. Rewriting Eq. (7.7) in the van der Pol form yields

$$\begin{cases} \dfrac{\mathrm{d}^2v_1}{\mathrm{d}t^2} + \omega_0^2v_1 = \omega_0^2v_1 + 2\left(\delta_0 - \delta_2v_1^2\right)\dfrac{\mathrm{d}v_1}{\mathrm{d}t} - \alpha_1\dfrac{\mathrm{d}v_2}{\mathrm{d}t} & (7.8a)\\ \dfrac{\mathrm{d}^2v_2}{\mathrm{d}t^2} + \omega_0^2v_2 = \omega_0^2v_2 + 2\left(\gamma_0 - \gamma_2v_2^2\right)\dfrac{\mathrm{d}v_2}{\mathrm{d}t} + \alpha_2\dfrac{\mathrm{d}v_1}{\mathrm{d}t} & (7.8b)\end{cases}$$

where ω_0 is the oscillation frequency, $\alpha_i = g_{mi}/(2C_i)$ is the ith stage coupling factor. Other values are the VDP parameters of the first stage are

$$\delta_0 = \frac{\frac{1}{2}g_{m0} - R_1^{-1}}{2C_1}, \qquad \delta_2 = \frac{3K^2}{2C_1 8 g_{m0}}, \tag{7.9}$$

and of the second stage are

$$\gamma_0 = \frac{\frac{1}{2}g_{m_0} - R_2^{-1}}{2C_2}, \qquad \gamma_2 = \frac{3K^2}{2C_2 8 g_{m0}}. \tag{7.10}$$

For the sinusoidal regime, the solution of Eq. (7.8) is of the form:

$$v_i = V_i \sin(\omega_0 t - \phi_i). \tag{7.11}$$

where V_i is the amplitude of the ith stage and ϕ_i is the phase. Using the harmonic balance method [9], the amplitude and phase transient equations are

$$\begin{cases} \frac{dV_1}{dt} = \delta_0 V_1 - \frac{\delta_2 V_1^3}{4} - \frac{\alpha_1 V_2}{2}\cos\Delta\phi & (7.12a)\\ \frac{dV_2}{dt} = \gamma_0 V_2 - \frac{\gamma_2 V_2^3}{4} + \frac{\alpha_2 V_1}{2}\cos\Delta\phi & (7.12b)\\ \frac{d\phi_1}{dt} = \frac{\omega_0}{2} - \alpha_1\frac{V_2}{2V_1}\sin\Delta\phi & (7.12c)\\ \frac{d\phi_2}{dt} = \frac{\omega_0}{2} - \alpha_2\frac{V_1}{2V_2}\sin\Delta\phi. & (7.12d) \end{cases}$$

where $\Delta\phi = \phi_2 - \phi_1$ represents the phase difference between the outputs.

Simplifying the system Eq. (7.12) assuming no mismatches between the stages, i.e. $\alpha_1 = \alpha_2 = \alpha$, $\delta_0 = \gamma_0$, and $\delta_2 = \gamma_2$, we obtain

$$\begin{cases} \frac{dV_1}{dt} = \delta_0 V_1 - \frac{1}{4}\delta_2 V_1^3 - \frac{\alpha}{2}V_2\cos\Delta\phi & (7.13a)\\ \frac{dV_2}{dt} = \delta_0 V_2 - \frac{1}{4}\delta_2 V_2^3 + \frac{\alpha}{2}V_1\cos\Delta\phi. & (7.13b)\\ \frac{d\phi_1}{dt} = \frac{\omega_0}{2} - \alpha\frac{V_2}{2V_1}\sin\Delta\phi & (7.13c)\\ \frac{d\phi_2}{dt} = \frac{\omega_0}{2} - \alpha\frac{V_1}{2V_2}\sin\Delta\phi. & (7.13d) \end{cases}$$

To determine the equilibrium points, we first subtract Eq. (7.13c) from Eq. (7.13d) to obtain

$$\frac{d\Delta\phi}{dt} = \frac{d\phi_2}{dt} - \frac{d\phi_1}{dt} = -\frac{\alpha}{2}\left(\frac{V_1}{V_2} - \frac{V_2}{V_1}\right)\sin\Delta\phi. \tag{7.14}$$

In steady state, $\mathrm{d}\Delta\phi/\mathrm{d}t = 0$, so

$$-\frac{\alpha}{2}\left(\frac{V_1}{V_2} - \frac{V_2}{V_1}\right)\sin\Delta\phi = 0. \tag{7.15}$$

We multiply both sides of Eq. (7.15) by V_1 and V_2, to avoid the indeterminate form 0/0 and obtain

$$-\frac{\alpha}{2}\left(V_1^2 - V_2^2\right)\sin\Delta\phi = 0. \tag{7.16}$$

From Eq. (7.16), one can see that the equilibrium points exist for $V_1 = V_2$ and for $\Delta\phi = \pm\pi$. Applying these conditions to Eqs. (7.13a) and (7.13b), we conclude that an equilibrium point, E_0, exists at $V_1 = V_2 = 0$. The third term on the right-hand sides of both Eqs. (7.13a) and (7.13b) has opposite signs. Hence, $\Delta\phi = \pm\pi/2$ and $V_1 = V_2$ also satisfy both equations. This leads us to the second equilibrium point: $E_1 = (V_{\mathrm{osc}}, V_{\mathrm{osc}}, \pi/2)$.

The equilibrium point E_0 has little interest because the oscillation amplitude is zero. We will derive the oscillation key parameters for the second equilibrium point, E_1. We assume equal output voltages, $V_1 = V_2 = V_{\mathrm{osc}}$, and quadrature outputs $\Delta\phi = \pi/2$. From Eq. (7.13a), or from Eq. (7.13b), we obtain the oscillation amplitude:

$$V_{\mathrm{osc}} = 2\sqrt{\frac{\delta_0}{\delta_2}} = \frac{8I_{\mathrm{level}}}{\sqrt{3}g_{m0}}\sqrt{\frac{Rg_{m0} - 2}{Rg_{m0}}}. \tag{7.17}$$

The oscillation frequency is obtained by adding Eqs. (7.13c) and (7.13d):

$$\omega_0 = \alpha = \frac{g_m}{2C}. \tag{7.18}$$

7.2.5 *Stability of the Equilibrium Points*

To understand how the circuit reaches the steady state, we do the stability analysis. In the previous section, we assumed equal amplitudes and perfect quadrature, since there are no mismatches between the stages. In this section, we will prove that this assumption is correct.

The voltage of the ith stage is given by Eq. (7.11). For $t = 0$, the voltage of the first stage is given by:

$$v_1(0) = V_1 \sin(-\phi_1). \tag{7.19}$$

The amplitude is not defined only by the initial conditions of the capacitance. Differentiating v_1 and dividing by ω, we obtain

$$\frac{1}{\omega}\frac{\mathrm{d}v_1}{\mathrm{d}t}\bigg|_{t=0} = V_1 \cos(-\phi_i). \tag{7.20}$$

Using $\omega = \alpha = g_m/(2C)$ in Eq. (7.20), we obtain

$$\frac{2}{g_m}C\frac{\mathrm{d}v_1}{\mathrm{d}t}\bigg|_{t=0} = \frac{i_c}{g_m/2} = V_1 \cos(-\phi_i). \tag{7.21}$$

The capacitor current, i_c, is the output current of the transconductance amplifier, i_1, plus the currents in the resistance, R, and in the negative resistance, R_N. We assumed that the negative resistance cancels R, resulting in:

$$i_c \approx i_1 = (g_m/2)v_2. \tag{7.22}$$

Substituting Eq. (7.22) into Eq. (7.21) results in:

$$v_2(0) = V_1 \cos(-\phi_i). \tag{7.23}$$

Combining Eq. (7.19) and Eq. (7.23) and solving with respect to the amplitude and phase, we obtain

$$V_1 = \sqrt{v_1^2(0) + v_2^2(0)}, \tag{7.24}$$

and

$$\phi_1 = -\tan^{-1}\left(\frac{v_1(0)}{v_2(0)}\right). \tag{7.25}$$

Similarly, for the second stage, we obtain

$$V_2 = \sqrt{v_1^2(0) + v_2^2(0)}, \tag{7.26}$$

and

$$\phi_2 = \tan^{-1}\left(\frac{v_2(0)}{v_1(0)}\right). \tag{7.27}$$

From Eqs. (7.24) and (7.26), we conclude that the amplitudes are equal, i.e. $V_1 = V_2 = V$. For the phase difference, we combine Eqs. (7.25) and (7.27) and use the mathematical identity ($\tan^{-1}(x) + \tan^{-1}(1/x) = \pi/2$), resulting in:

$$\Delta\phi = \phi_2 - \phi_1 = \tan^{-1}\left(\frac{v_1(0)}{v_2(0)}\right) + \tan^{-1}\left(\frac{v_2(0)}{v_1(0)}\right) = \frac{\pi}{2}. \tag{7.28}$$

Simplifying the system Eq. (7.13) by combining Eqs. (7.13c) and (7.13d) and using $V_1 = V_2 = V$ and $\Delta\phi = \pi/2$, we obtain

$$\begin{cases} \dfrac{\mathrm{d}V}{\mathrm{d}t} = \delta_0 V - \dfrac{1}{4}\delta_2 V^3 & (7.29\text{a}) \\ \dfrac{\mathrm{d}\Delta\phi}{\mathrm{d}t} = 0. & (7.29\text{b}) \end{cases}$$

The system in equation (7.29) is equivalent to the system obtained for the VDPO and it was already solved in Sect. 3.2. Thus, we conclude that the equilibrium point $E_0 = (0, 0, \Delta\phi)$ is unstable and the equilibrium point $E_1 = (V_{\text{osc}}, V_{\text{osc}}, \pi/2)$ is stable.

7.2.6 Oscillator with Mismatches

Now, we determine the impact of the components' mismatch on the key oscillator parameters. First, the steady-state solutions (equilibrium points) of the system equation (7.12) are calculated by equating all derivatives to zero (i.e. $\mathrm{d}V_{o1}/\mathrm{d}t = \mathrm{d}V_{o2}/\mathrm{d}t = 0$ and $\mathrm{d}\phi_1/\mathrm{d}t = \mathrm{d}\phi_2/\mathrm{d}t = 0$). This gives

$$\begin{cases} \delta_0 - \dfrac{\delta_2 V_1^2}{4} - \dfrac{\alpha_1}{2}\dfrac{V_2}{V_1}\cos(\Delta\phi) = 0 & (7.30\text{a}) \\ \gamma_0 - \dfrac{\gamma_2 V_2^2}{4} + \dfrac{\alpha_2}{2}\dfrac{V_1}{V_2}\cos(\Delta\phi) = 0 & (7.30\text{b}) \\ \dfrac{\omega_0}{2} - \dfrac{\alpha_1}{2}\dfrac{V_2}{V_1}\sin(\Delta\phi) = 0 & (7.30\text{c}) \\ \dfrac{\omega_0}{2} - \dfrac{\alpha_2}{2}\dfrac{V_1}{V_2}\sin(\Delta\phi) = 0 & (7.30\text{d}) \end{cases}$$

Now, Eq. (7.31a) and Eq. (7.31b) are obtained by adding and subtracting, respectively, Eq. (7.30a) and Eq. (7.30b). Similarly, Eq. (7.31c) and Eq. (7.31d) are obtained by subtracting and adding, respectively, Eq. (7.30c) and Eq. (7.30d):

$$\begin{cases} \delta_0 + \gamma_0 - \dfrac{\delta_2 V_1^2 + \gamma_2 V_2^2}{4} = \dfrac{1}{4}\left(\alpha_1 \dfrac{V_2}{V_1} - \alpha_2 \dfrac{V_1}{V_2}\right)\cos(\Delta\phi) & \text{(7.31a)} \\ \delta_0 - \gamma_0 - \dfrac{\delta_2 V_1^2 - \gamma_2 V_2^2}{4} = \dfrac{1}{4}\left(\alpha_1 \dfrac{V_2}{V_1} + \alpha_2 \dfrac{V_1}{V_2}\right)\cos(\Delta\phi) & \text{(7.31b)} \\ \left(\alpha_2 \dfrac{V_1}{4V_2} - \alpha_1 \dfrac{V_2}{4V_1}\right)\sin(-\Delta\phi) = 0. & \text{(7.31c)} \\ \omega_0 = \left(\alpha_1 \dfrac{V_2}{4V_1} + \alpha_2 \dfrac{V_1}{4V_2}\right)\sin(\Delta\phi). & \text{(7.31d)} \end{cases}$$

Assuming that the outputs are nearly in quadrature, i.e. $\Delta\phi \approx \pm\pi/2$ and using the approximation ($\sin\Delta\phi \approx 1$), Eq. (7.31c) is reduced to:

$$\frac{1}{2}\left(\alpha_2 \frac{V_1}{V_2} - \alpha_1 \frac{V_2}{V_1}\right) = 0, \tag{7.32}$$

This leads us to the conclusion that the relationship between amplitudes is

$$V_1 = \sqrt{\frac{\alpha_1}{\alpha_2}}V_2 = \sqrt{\frac{g_{m1}C_2}{g_{m2}C_1}}V_2. \tag{7.33}$$

A simplified equation for the oscillation frequency can be obtained by substituting Eq. (7.33) into Eq. (7.31) and assuming the outputs nearly in quadrature again, i.e. $\Delta\phi \approx \pi/2$. This results in:

$$\omega_0 \approx \frac{1}{2}\sqrt{\alpha_1\alpha_2} = \frac{1}{2}\sqrt{\frac{g_{m1}g_{m2}}{C_1C_2}}, \tag{7.34}$$

Substituting $g_{m1} = g_m\left(1 - \Delta g_m/(2g_m)\right)$, $g_{m2} = g_m\left(1 + \Delta g_m/(2g_m)\right)$, $C_1 = C\left(1 - \Delta C/(2C)\right)$, and $C_2 = C\left(1 + \Delta C/(2C)\right)$ into the right-hand side of Eq. (7.34) gives

$$\omega_0 \approx \frac{g_m}{2C}\sqrt{\frac{1 - \left(\frac{\Delta g_m}{2g_m}\right)^2}{1 - \left(\frac{\Delta C}{2C}\right)^2}}, \tag{7.35}$$

where $\Delta g_m = g_{m2} - g_{m1}$ and $\Delta C = C_2 - C_1$. For small mismatches, $\left(\frac{\Delta g_m}{2g_m}\right)^2 \ll 1$ and $\left(\frac{\Delta C}{2C}\right)^2 \ll 1$, and Eq. (7.35) becomes

$$\omega_0 \approx \frac{g_m}{2C}. \tag{7.36}$$

To derive the amplitude mismatch, we use the definition:

$$\epsilon_A = 2\frac{V_2 - V_1}{V_2 + V_1} = 2\frac{1 - \sqrt{\frac{\alpha_1}{\alpha_2}}}{1 + \sqrt{\frac{\alpha_1}{\alpha_2}}}. \tag{7.37}$$

If the coupling factors, α_i, in Eq. (7.37) are equal, then perfect amplitude match is obtained. Using in Eq. (7.37) the coupling factors $\alpha_i = g_{mi}/(2C_i)$ and the circuit parameters as function of the mismatches and rearranging the terms, we obtain

$$\epsilon_A = 2\frac{\sqrt{1 + \frac{\Delta g_m}{2g_m} - \frac{\Delta C}{2C} - \frac{\Delta g_m}{2g_m}\frac{\Delta C}{2C}} - \sqrt{1 - \frac{\Delta g_m}{2g_m} + \frac{\Delta C}{2C} - \frac{\Delta g_m}{2g_m}\frac{\Delta C}{2C}}}{\sqrt{1 + \frac{\Delta g_m}{2g_m} - \frac{\Delta C}{2C} - \frac{\Delta g_m}{2g_m}\frac{\Delta C}{2C}} + \sqrt{1 - \frac{\Delta g_m}{2g_m} + \frac{\Delta C}{2C} - \frac{\Delta g_m}{2g_m}\frac{\Delta C}{2C}}}. \tag{7.38}$$

If the mismatches are small, the terms $\left(\frac{\Delta g_m}{2g_m}\frac{\Delta C}{2C}\right)$ in Eq. (7.38) can be omitted. Moreover, using the approximation $\sqrt{1 \pm x} \approx 1 \pm x/2$ for $|x| \ll 1$ we find that

$$\epsilon_A \approx \frac{1}{2}\left[\left(\frac{\Delta g_m}{g_m}\right) - \left(\frac{\Delta C}{C}\right)\right], \tag{7.39}$$

Hence, the amplitude mismatch depends only on the transconductance and capacitance mismatches. The capacitance mismatches can be minimized by a careful layout, but cannot be fully eliminated. The transconductances' mismatches can be controlled by adjusting the tail currents of the source-coupled pairs. By controlling each transconductance independently so that the transconductance mismatch is equal to the capacitance mismatch results in a perfect amplitude matching, i.e. $\epsilon_A = 0$.

The amplitude value without mismatch and quadrature condition is given by the VDPO amplitude equation. It can be obtained from Eq. (7.30a) that

$$V_{\text{osc}} = 2\sqrt{\frac{\delta_0}{\delta_2}} = \frac{4}{\sqrt{3}K}\sqrt{\frac{g_{m0}}{R}(Rg_{m0} - 2)} = \frac{8I_{level}}{\sqrt{3}g_{m0}}\sqrt{\frac{Rg_{m0} - 2}{Rg_{m0}}}. \tag{7.40}$$

where I_{level} is the tail current of the negative-resistance circuit. If the mismatches are present, the amplitudes are given by:

$$V_1 = V_{\text{osc}}\left(1 - \frac{1}{2}\epsilon_A\right) = \frac{2I_{\text{level}}}{\sqrt{3}g_{m0}}\sqrt{\frac{Rg_{m0} - 2}{Rg_{m0}}}\left[4 - \left(\frac{\Delta g_m}{g_m}\right) + \left(\frac{\Delta C}{C}\right)\right]. \tag{7.41}$$

and

$$V_2 = V_{\text{osc}}\left(1 + \frac{1}{2}\epsilon_A\right) = \frac{2I_{\text{level}}}{3\sqrt{3}g_{m0}{}^2}\sqrt{\frac{Rg_{m0} - 2}{Rg_{m0}}}\left[4 + \left(\frac{\Delta g_m}{g_m}\right) - \left(\frac{\Delta C}{C}\right)\right] \tag{7.42}$$

As will be shown next, the amplitude matching reduces the phase error.

To derive the phase error, we divide Eq. (7.31d) by Eq. (7.31b). This gives

$$\tan(\Delta\phi) = \frac{\omega_0}{(\delta_0 - \gamma_0) - \frac{\delta_2 V_1^2 - \gamma_2 V_2^2}{4}}. \tag{7.43}$$

To obtain from Eq. (7.43) the phase error, ϵ_ϕ, we relate the phase difference to the phase error as $\Delta\phi = \pi/2 - \epsilon_\phi$. Then, using the trigonometric identity $\tan\Delta\phi = \cot\epsilon_\phi$, we obtain

$$\tan(\epsilon_\phi) = \frac{1}{\cot\epsilon_\phi} = \frac{4(\delta_0 - \gamma_0) - \left(\delta_2 V_1^2 - \gamma_2 V_2^2\right)}{4\omega_0}. \tag{7.44}$$

For small mismatches, $\epsilon_\phi \ll \pi/2$, the Taylor series approximation gives

$$\epsilon_\phi \approx \frac{4(\delta_0 - \gamma_0) - (\delta_2 V_1^2 - \gamma_2 V_2^2)}{4\omega_0}. \tag{7.45}$$

From the definition of VDP parameters, Eqs. (7.9) and (7.10), it follows that:

$$\delta_0 - \gamma_0 = \frac{g_{m0}R_1 - 2}{4R_1C_1} - \frac{g_{m0}R_2 - 2}{4R_2C_2} \tag{7.46}$$

Substituting $R_1 = R(1 - \Delta R/(2R))$, $R_2 = R(1 + \Delta R/(2R))$, $C_1 = C(1 - \Delta C/(2C))$ and $C_2 = C(1 + \Delta C/(2C))$ into Eq. (7.46) and rearranging the terms, we obtain

$$\delta_0 - \gamma_0 = \frac{-2RC\left(\frac{\Delta R}{R} + \frac{\Delta C}{C}\right) + g_{m0}R^2C\left(\frac{\Delta C}{C}\right)}{4R^2C^2\left[1 - \left(\frac{\Delta R}{2R}\right)^2\right]\left[1 + \left(\frac{\Delta C}{2C}\right)^2\right]}. \tag{7.47}$$

where $\Delta R = R_2 - R_1$ and $\Delta C = C_2 - C_1$. Assuming small mismatches, Eq. (7.47) can be reduced to:

$$\delta_0 - \gamma_0 \approx -\frac{2\left(\frac{\Delta R}{R}\right) - (g_{m0}R - 2)\left(\frac{\Delta C}{C}\right)}{4RC}. \tag{7.48}$$

Now, substituting the VDP parameters, Eqs. (7.9) and (7.10), in the second term of Eq. (7.48), $\delta_2 V_1^2 - \gamma_2 V_2^2$, results in:

$$\delta_2 V_1^2 - \gamma_2 V_2^2 = \frac{3K^2}{16g_{m0}} \left(\frac{V_1^2}{C_1} - \frac{V_2^2}{C_2} \right) \tag{7.49}$$

Using again $R_1 = R\,(1 - \Delta R/(2R))$, $R_2 = R\,(1 + \Delta R/(2R))$, $C_1 = C\,(1 - \Delta C/(2C))$ and $C_2 = C\,(1 + \Delta C/(2C))$ in Eq. (7.49) and rearranging the terms, we obtain

$$\delta_2 V_1^2 - \gamma_2 V_2^2 = \frac{3K^2}{16g_{m0}} \frac{\left(V_1^2 - V_2^2\right) + \left(\frac{\Delta C}{2C}\right)\left(V_1^2 + V_2^2\right)}{C\left[1 - \left(\frac{\Delta C}{2C}\right)^2\right]} \tag{7.50}$$

Now, we substitute $V_1 = V_{\text{osc}}\,(1 - \epsilon_A/2)$ and $V_2 = V_{\text{osc}}\,(1 + \epsilon_A/2)$ into Eq. (7.50). Assuming $\left(\frac{\Delta C}{2C}\right)^2 \ll 1$, the result is

$$\delta_2 V_1^2 - \gamma_2 V_2^2 \approx \frac{3K^2}{16g_{m0}C} V_{\text{osc}}^2 \left(\frac{\Delta C}{C} - 2\epsilon_A \right). \tag{7.51}$$

Further, substituting Eq. (7.39) and Eq. (7.40) into Eq. (7.51) gives

$$\delta_2 V_1^2 - \gamma_2 V_2^2 \approx 2\frac{Rg_{m0} - 2}{RC} \left[2\frac{\Delta C}{C} - \left(\frac{\Delta g_m}{g_m} \right) \right]. \tag{7.52}$$

Finally, substituting Eq. (7.36), Eq. (7.47) and Eq. (7.52) into Eq. (7.45), we obtain

$$\epsilon_\phi \approx -\left[\left(\frac{\Delta R}{R} \right) + \left(\frac{\Delta C}{C} - \frac{\Delta g_m}{g_m} \right) \frac{(Rg_{m0} - 2)}{2} \right] \frac{1}{Rg_m}. \tag{7.53}$$

Equation (7.53) gives an interesting insight: one can reduce the quadrature phase error by increasing the amplifier gain and, also, by equalizing the capacitance and the amplifier's transconductance mismatches. Moreover, if an imbalance between the capacitances' and transconductances' mismatches compensates the resistances' mismatch, a zero phase error is obtained. Note that the units in Eq. (7.53) are radians. To obtain the result in degrees, we multiply Eq. (7.53) by $180/\pi$ which gives

$$\epsilon_\phi \approx -\frac{180}{\pi} \left[\left(\frac{\Delta R}{R} \right) + \left(\frac{\Delta C}{C} - \frac{\Delta g_m}{g_m} \right) \frac{(Rg_{m0} - 2)}{2} \right] \frac{1}{Rg_m}. \tag{7.54}$$

7.3 Simulation Results

The circuit of Fig. 7.2 was simulated using the parameters of a 130 nm standard CMOS technology. The oscillation frequency is 2.4 GHz. The circuit components are $C_1 = C_2 = C$=77 fF, $R_1 = R_2 = R = 600\,\Omega$, $(W/L) = 115.2\,\mu\text{m}/120\,\text{nm}$ for

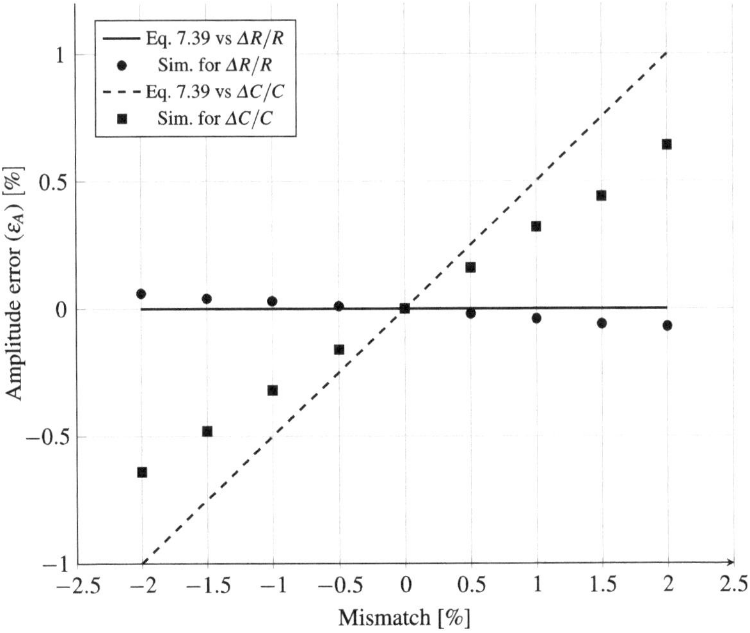

Fig. 7.8 Impact of the mismatches on the amplitude error

transistors M_1, M_2, M_5, and M_6, $(W/L) = 14.4\,\mu\text{m}/120\,\text{nm}$ for M_3, M_4, M_7, and M_8, $I_{\text{level}} = 0.8\,\text{mA}$, $I = 2\,\text{mA}$, and the supply voltage is 1.2 V. The voltage and current sources are assumed to be ideal.

To validate Eq. (7.39), we run several simulations with $\frac{\Delta C}{C} = 0$ and $\frac{\Delta g_m}{g_m} = 0$ and sweep the resistance mismatch ΔR from -2% to $+2\%$. Although, in the theory (Eq. 7.39) the amplitude error does not depend on the resistors' mismatch, the results show that, indeed, the resistance mismatches have only 0.1% as maximum contribution to the amplitude error (Fig. 7.8). Thus, the impact of the resistance mismatch on the amplitude error is negligible.

To further validate the amplitude mismatch given by Eq. (7.39), several simulations were made varying the mismatches between the capacitances, C_1 and C_2. The results show a significant impact on the amplitude mismatch, as shown in Fig. 7.8. The deviation between simulation and analytical results are explained by the nonlinearities of the transistors' capacitances and the parasitic capacitances.

The phase error Eq. (7.54) was also validated by simulation. First, we simulate the circuit with matched capacitances and a sweep of the resistance mismatches. There is a small discrepancy between simulation data and the theory, as shown in Fig. 7.9. However, we can conclude that with the increase of the transconductance the phase error decreases. Afterwards, we simulate the circuit with only a capacitance mismatch of 2%. Again, the phase error decreases with the increase of the amplifier's transconductances, which is in line with the trend indicated by Eq. (7.54).

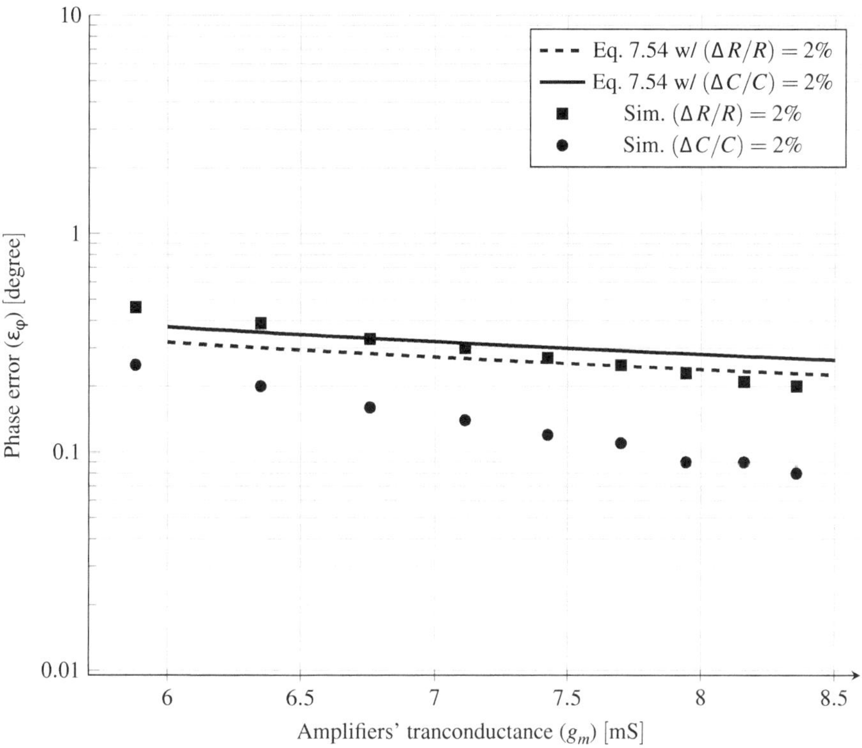

Fig. 7.9 Phase error as function of transconductance

To determine the impact of the amplifier's transconductance, g_m, on the phase error and phase noise, the circuit was simulated with a constant component mismatch of 2% and increasing g_m. Two results are shown in Fig. 7.10 where the phase noise, at the offset of 10 MHz, is represented by white circles. The results show that the phase error is reduced when g_m increases, which is in agreement with the trend described by Eq. (7.54). To compensate the frequency shift, capacitance C was adjusted in each simulation to maintain the oscillation frequency close to 2.4 GHz.

To compare this oscillator with other works, we use the conventional figure-of-merit (FoM) [11]:

$$\text{FoM} = \text{PN} + 10\log\left(\left(\frac{\Delta f}{f}\right)^2 \frac{P_{DC}}{P_{ref}}\right), \tag{7.55}$$

where PN is the phase noise, P_{DC} is the dissipated power in mW, P_{ref} is the reference power of 1 mW, f is the oscillation frequency and Δf is the frequency offset.

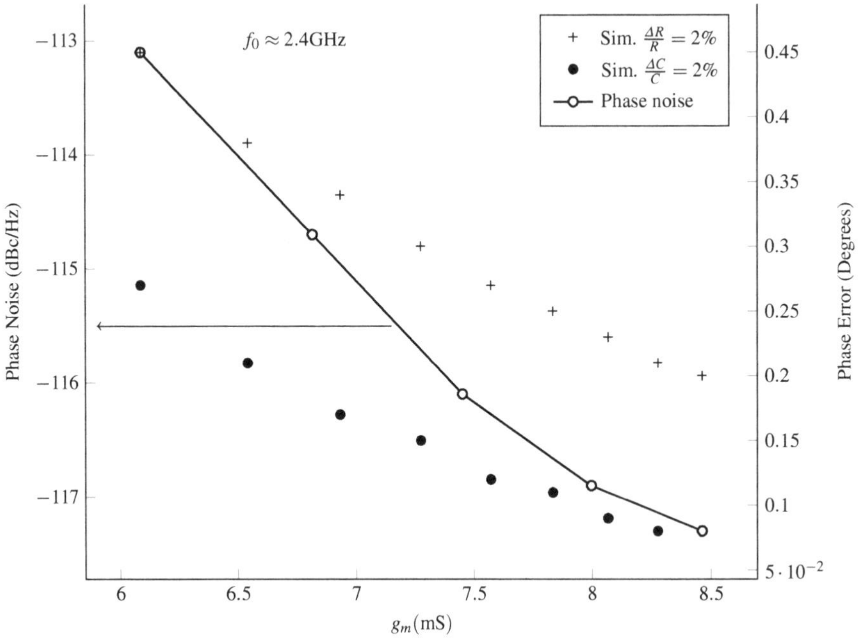

Fig. 7.10 Phase noise and phase error as function of the transconductance [10]

Table 7.1 Comparison of state-of-the-art nearly sinusoidal RC-oscillators with a similar circuit topology

Reference	f (GHz)	PN (dBc/Hz)	Δf (MHz)	P_{DC} (mW)	FoM	IQ	$\frac{f_{max}}{f_{min}}$
[12]	9.8	−94	2	75	−149.1	Yes	1.17
[13]	1.4	−117.3	10	9.6	−150.4	Yes	2.33
[14]	2.5	−95.4	1	2.8	−158.9	No	1.22
[15]	3.1	−110.3	10	7.7	−151.2	Yes	3.42
This work	2.4	−115	10	6.7	−153.2	Yes	5.47

Table 7.1 gives a comparison among the state-of-the-art of inductorless quadrature oscillators. A figure of merit (FoM) of −153.2 dBc/Hz is obtained for a power of 6.72 mW, which is the best performance for a QVCO with nearly sinusoidal output.

7.4 Conclusions

We presented the analysis of the low-power and wide tuning range quadrature oscillator using the configuration with two integrators. The analysis is focused on the amplitude and phase error as functions of the component mismatches. To minimize the impact of the mismatches on the quadrature error, the designer should increase the amplifier's transconductances. Increasing the transconductances also reduces the phase noise, unlike what happens in LC-oscillators, which have a trade-off between phase noise and phase error. The quadrature error can be minimized, and in some cases eliminated, by adjusting the transconductances to compensate the capacitance mismatch. However, to obtain outputs in perfect quadrature, we must allow some amplitude error. Also, to limit the circuit mismatches, passive components can be replaced by their MOSFET counterparts, which due to the low process variations have less relative mismatches [16]. Furthermore, this allows the circuit to be trimmed.

An automatic compensation of the mismatches may be performed by an auxiliary control circuit. This control circuit should do two independent adjustments. First, to minimize the amplitude mismatch, the current sources I_{level} should be adjusted based on the difference between the amplitudes of both stages. Then, the resistance values can be adjusted based on the difference between the common-mode voltages of the two outputs.

References

1. L. Oliveira, J. Fernandes, I. Filanovsky, C. Verhoeven, M. Silva, *Analysis and Design of Quadrature Oscillators* (Springer, Heidelberg, 2008)
2. B. Razavi, *RF Microelectronics* (Prentice-Hall, Upper Saddle River, 1998)
3. P.-I. Mak, S.-P. U, R. Martins, Transceiver architecture selection: review, state-of-the-art survey and case study. IEEE Circuits Syst. Mag. **7**(2), 6–25 (2007)
4. L. Oliveira, J. Fernandes, I. Filanovsky, C.J.M. Verhoeven, Wideband two-integrator oscillator-mixer, in *6th International Conference On ASIC (ASICON)*, vol. 1 (2005), pp. 642–645
5. J. van der Tang, D. Kasperkovitz, A 0.9-2.2 GHz monolithic quadrature mixer oscillator for direct-conversion satellite receivers, in *43rd IEEE International Solid-State Circuits Conference* (1997), pp. 88–89
6. J. Casaleiro, L. Oliveira, I. Filanovsky, Quadrature error of two-integrator oscillators, in *IEEE 57th International Midwest Symposium on Circuits and Systems (MWSCAS)* (2014), pp. 483–486
7. I.M. Filanovsky, C.J. Verhoeven, Sinusoidal and relaxation oscillations in Source-Coupled multivibrators. IEEE Trans. Circuits Syst. II **54**(11), 1009–1013 (2007)
8. J. Casaleiro, H.F. Lopes, L.B. Oliveira, I. Filanovsky, Analysis and design of CMOS coupled multivibrators. Int. J. Microelectron. Comput. Sci. **1**, 249–256 (2010)
9. D.A. Johns, K. Martin, *Analog Integrated Circuit Design*, 2nd edn. (Wiley, New York, 2011)
10. J. Casaleiro, L.B. Oliveira, I.M. Filanovsky, Amplitude and quadrature errors of two-integrator oscillator. J. Low Power Electron. **11**, 340–348 (2015)
11. J.O. Plouchart, H. Ainspan, M. Soyuer, A. Ruehli, A fully-monolithic SiGe differential voltage-controlled oscillator for 5 GHz wireless applications, in *IEEE Radio Frequency Integrated Circuits (RFIC) Symp. - Dig. Papers* (2000), pp. 57–60
12. J. van der Tang, D. Kasperkovitz, A. van Roermund, A 9.8-11.5-ghz quadrature ring oscillator for optical receivers. IEEE J. Solid State Circuits **37**(3), 438–442 (2002)
13. E. Ortigueira, J. Fernandes, M. Silva, L. Oliveira, An ultra compact wideband combined LNA-oscillator-mixer for biomedical applications, in *IEEE 55th International Midwest Symposium on Circuits and Systems (MWSCAS)* (2012), pp. 162–165

14. S.W. Park, E. Sanchez-Sinencio, RF oscillator based on a passive RC bandpass filter. IEEE J. Solid State Circuits **44**(11), 3092–3101 (2009)
15. J. Fernandes, L. Oliveira, E. Snelling, M. Silva, An inductorless CMOS quadrature oscillator continuously tuneable from 3.1 to 10.6 GHz. Int. J. Circuit Theory Appl. **40**, 209–216 (2012)
16. Y. Cheng, The influence and modeling of process variation and device mismatch for analog/RF circuit design, in *Proceedings of the Fourth IEEE International Caracas Conference on Devices, Circuits and Systems* (2002), pp. D046–1–D046–8

Chapter 8
Conclusions and Suggested Further Research

8.1 Conclusions

Quadrature oscillators (QO) are key blocks in modern receivers, e.g. in the low-IF and zero-IF receivers. These receivers allow full integration. This reduces the overall cost, but their performance is directly related to the image rejection ratio, which in turn depends on the amplitude and quadrature errors of the QO. The objective of the study reported in this book was to investigate the impact of the components' mismatches on the amplitude and phase errors of quadrature RC-oscillators working in the quasi-sinusoidal regime. Three quadrature oscillators were investigated: the active coupling RC-oscillator (in Chap. 5), the capacitive coupling RC-oscillator (in Chap. 6) and the two-integrator oscillator (in Chap. 7). This study showed that amplitude and quadrature errors are directly proportional to the components' mismatches and inversely proportional to the coupling factor. Thus, for typical mismatches (around 1%) with a standard 130-nm CMOS technology, it is possible to design a quadrature RC-oscillator with amplitude error below 1% and quadrature error below 1°.

Before the analysis of coupled oscillators, it was shown in Chap. 3 that the single RC-oscillator working in the sinusoidal regime can be modeled by the van der Pol equation. The circuit nonlinearities are similar to the nonlinear term in the van der Pol VDPO. Thus, we used in this book the VDPO as a model for the study of coupled oscillators. Simulation results showed that the theoretical model predicts well the oscillation amplitude. For the oscillation frequency, however, the simulation results showed that the oscillation frequency is 10% above the value given by the theory. This difference can be explained by the approximations and, especially, by the effect of the channel-length modulation in the transistors (which was not accounted for in the model). This factor increases the transconductances of the transistors, leading to an increase of the oscillation frequency. The approximation of the single RC-oscillator by the VDPO was already used in other research works, but without relating the transistors' parameters modes to the VDP parameters.

J. C. F. d. A. Casaleiro et al., *Quadrature RC–Oscillators*, Analog Circuits and Signal Processing, https://doi.org/10.1007/978-3-030-00740-9_8

Coupling of two oscillators can be considered as injecting a signal from one oscillator into the other. Hence, we study first the injection of a sinusoidal signal into a single RC-oscillator (injection locking). After that, we substitute the external source by a second oscillator and derive the results for coupled oscillators. From the injection locking study (Chap. 4), we conclude that an oscillator locks to the frequency of the injected signal within a limited band (locking range). The phase difference between the two signals and the oscillation amplitude are adjusted by the attenuation and phase imposed by the oscillator's resonant tank at the locked frequency. Moreover, the locking range is inversely proportional to the resonant tank quality factor and directly proportional to the ratio between the injected current and the oscillator's current amplitudes. Thus, oscillators with high-Q resonant tanks are more difficult to couple, and they have worse performance than oscillators with low-Q resonant tanks. This is not a problem for coupling of RC-oscillators, since the maximum quality factor of these oscillators is one.

The study of the active and capacitive coupling showed that both methods are viable solutions to generate quadrature outputs. Both coupling methods showed similar results for the amplitude and quadrature errors. The error equations were derived with respect to the resistance and capacitance mismatches. It was found that the amplitude error is proportional to both mismatches and is inversely proportional to the coupling factor.

In the active coupling (Chap. 5), the effects of the mismatches are independent and, therefore, are cumulative. The impact of the resistance mismatches is similar to the capacitive mismatch. The coupling factor can be used to decrease the amplitude error and phase error. The impact of the resistance mismatches is slightly higher than that of the capacitance mismatches. It was also found that the oscillation frequency is insensitive to the mismatches, but decreases with the increase in the coupling factor.

In the capacitive coupling (Chap. 6), the effects of the mismatches are also independent and, therefore, are cumulative. The impact of the resistance mismatches is slightly different. There is a constant term (not dependent on the coupling factor) that sets the minimum value for the amplitude error. The coupling factor can be used to decrease the amplitude error. The phase error depends on the resistance and capacitance mismatches and is inversely proportional to the coupling factor. The impact of the resistance mismatches is slightly higher than that of the capacitance mismatches. Moreover, it was found that the phase error is proportional to the amplitude error. It was also found that the increase in the coupling factor has almost no impact on the phase noise.

The two-integrator oscillator (Chap. 7) has a working principle fundamentally different from that of coupled oscillators. Although being an RC-oscillator (inductorless), it has a single-loop topology, with inherent quadrature outputs. For the two-integrator oscillator, the error equations were derived with respect to the resistance, capacitance and transconductance mismatches. Similarly to the other two quadrature oscillators, the phase error is directly proportional to the resistance and capacitance mismatches and inversely proportional to the transconductances (which have a similar role to the coupling factor). The transconductance mismatches, however, may compensate the capacitive mismatches. Hence, if the transconductances are adjusted such that their mismatch compensates both the resistance and capacitance mismatches, a perfect quadrature can be obtained. The amplitude error depends mainly on the transconductance and capacitance mismatches. However, these mismatches have an opposite impact on the amplitude error. The transconductance mismatch increases the error and the capacitance mismatch decreases it. Although, in theory, the amplitude error does not depend on the resistance mismatch, simulations showed that there is a weak dependence and, therefore, to obtain outputs in perfect quadrature one must allow some amplitude error. This is due to the channel-length modulation effect. Further, the simulation showed that the phase noise decreases with the increase of the transconductance. Thus, contrarily to what happens in coupled LC-oscillators, there is no trade-off between the phase noise and phase error.

8.2 Further Research Topics

The following points might be interesting as the subject of future research.

- For the capacitive coupling oscillator, only the oscillation frequency was validated by measurement results and the amplitude- and phase-error equations have been validated only by simulation. The validation of the theory by measurements on a prototype circuit is desirable.
- For the active coupling oscillator and the two-integrator oscillator, the oscillation frequency, amplitude- and phase-error equations were validated by simulation only. The validation of the theory by measurements on a prototype circuit is desirable.
- In the study of the two-integrator oscillator, it was found that the transconductance mismatch counterbalances the resistance and capacitance mismatches. This is not the case for other oscillators. This suggests the possibility of substantially reducing the quadrature error. An important topic of future research is to investigate the possibility to design a circuit to automatically compensate these mismatches. This control circuit should independently adjust the transconductances to compensate the resistance and capacitance mismatches. This is a challenging task since the compensation circuit also has its mismatches.
- Prototype circuits implemented in 130 nm standard CMOS were used to validate the equations. Verifying the validity of the theory presented in this book for more recent technologies might be of interest.

In this book, it was shown that RC-oscillators are a viable alternative to LC-oscillators. Although the performance is inferior than that of LC-oscillators, for low-power and low-cost applications RC-oscillators are a good choice. Applications like wide area sensor networks for Smart-Cities and Internet-of-Things (IoT) will require a large number of small form-factor and low-power devices. Also, the necessity of low-cost devices forces the development of standards with more relaxed specifications for the transceivers; examples are the Bluetooth Low Energy (BLE) standard, or the LoRa, a Low-Power Wide Area Network for IoT. Thus, in the authors' opinion the RC-oscillators are a suitable choice for these devices, and, it is expected that in the near future the use of RC-oscillators will increase.

Appendix A
Van der Pol Oscillator

A.1 Forced Oscillations in a Linear Conservative System

Let us consider a linear one degree of freedom conservative system described by the differential equation:

$$\frac{d^2v}{dt^2} + \omega^2 v = F_0 \cos pt. \tag{A.1}$$

The solution of Eq. (A.1) includes a free and a forced component. Calculating them, one finds that the full solution of Eq. (A.1) can be written as:

$$v = a \sin \omega t - b \cos \omega t + \frac{F_0}{\omega^2 - p^2} \cos pt. \tag{A.2}$$

If the system initial conditions are:

$$v|_{t=0} = \frac{dv}{dt}|_{t=0} = 0. \tag{A.3}$$

then the solution becomes

$$v = \frac{F_0}{\omega^2 - p^2} (\cos pt - \cos \omega t) = \frac{F_0}{\omega + p} t \frac{\cos pt - \cos \omega t}{(\omega - p)\, t}. \tag{A.4}$$

If the frequency p of external force is approaching to the frequency ω of the free oscillations, one arrives to the resonance, and the solution of Eq. (A.4) becomes

$$\lim_{p \to \omega} v = \frac{F_0}{2\omega} t \sin \omega t. \tag{A.5}$$

J. C. F. d. A. Casaleiro et al., *Quadrature RC–Oscillators*, Analog Circuits and Signal Processing, https://doi.org/10.1007/978-3-030-00740-9

Hence, one can say that the external harmonic excitation at the resonance frequency results in harmonic oscillations with increasing amplitude. The phase of these oscillations is shifted by a quarter period with respect to the excitation. Or, we can say that the cosinusoidal external excitation of the resonant frequency results in sinusoidal oscillations with linearly increasing amplitude.

In a similar way, one can obtain that the sinusoidal external excitation of the resonant frequency will develop cosinusoidal oscillations the amplitude of which is linearly increasing with time. Indeed, one can find that, with the same initial conditions, the equation:

$$\frac{d^2v}{dt^2} + \omega^2 v = f_0 \sin pt. \tag{A.6}$$

has the full solution:

$$v = -\frac{f_0}{2\omega} t \cos \omega t. \tag{A.7}$$

(notice the negative sign). But, if

$$\frac{d^2v}{dt^2} + \omega^2 v = F_0 \cos \omega t. \tag{A.8}$$

the full solution will be

$$v = \frac{F_0}{2\omega} t \sin \omega t. \tag{A.9}$$

Finally, when the external excitation has both cosinusoidal and sinusoidal components (of the resonant frequency):

$$\frac{d^2v}{dt^2} + \omega^2 v = F_0 \cos \omega t + f_0 \sin \omega t. \tag{A.10}$$

then the full solution will be

$$v = \frac{F_0}{2\omega} t \sin \omega t - \frac{f_0}{2\omega} t \cos \omega t. \tag{A.11}$$

This solution can be rewritten as:

$$v = a(t) \sin \omega t + b(t) \cos \omega t. \tag{A.12}$$

The amplitudes of harmonic components of this solution are increasing with time, at a rate given by:

$$\frac{da}{dt} = \frac{F_0}{2\omega} \tag{A.13}$$

for the sinusoidal component and

$$\frac{db}{dt} = \frac{f_0}{2\omega} \tag{A.14}$$

for the cosinusoidal component.

A.2 Theodorchik's Method

K.F. Theodorchik considers the oscillating systems described by the nonlinear second-order differential equation:

$$\frac{d^2v}{dt^2} + 2\delta_0\psi\left(v, \frac{dv}{dt}\right)\frac{dv}{dt} + \omega_0^2 v = 0 \tag{A.15}$$

The steady-state solution of this equation will be nearly sinusoidal if the condition

$$|\epsilon| = \frac{|\delta_0|}{\omega_0} \ll 1 \tag{A.16}$$

is satisfied and the nonlinear function is limited, that is:

$$-N \leq \psi\left(v, \frac{dv}{dt}\right) \leq N \tag{A.17}$$

Here, N is an arbitrary number. The condition Eq. (A.17) is always satisfied in transistor circuits. These two conditions provide also a nearly sinusoidal form for each oscillation period during the transients, when the oscillator is turned ON (or, in other words that the oscillation amplitude increases rather slowly).

Then, one can try to find the solution of Eq. (A.15) as:

$$v(t) = a(t)\sin\omega t \tag{A.18}$$

where the amplitude $a(t)$ changes very slowly and ω is constant. This means that the time θ elapsing during transient is large in comparison with the oscillation period, i.e. $\theta \gg T = (2\pi)/\omega$ and in the interval Δt which is small in comparison with θ but large in comparison with T, i.e. for $\theta \gg \Delta t \gg T$ one can neglect the variation of the oscillation amplitude. Hence, for the interval Δt the solution can be approximated as:

$$v(t) = a\sin\omega t \tag{A.19}$$

where a and ω are constant.

Then, to find the values of ω and the equation for the amplitude transient, Eq. (A.15) is rewritten as:

$$\frac{d^2v}{dt^2} + \omega_0^2 v = \left(\omega^2 - \omega_0^2\right) - 2\delta_0 \psi\left(v, \frac{dv}{dt}\right)\frac{dv}{dt} \tag{A.20}$$

Let us consider the right-hand side of Eq. (A.20) as the sum of inner forces of the system applied to the conservative oscillating system the equation of which is written on the left-hand side of Eq. (A.20). Then, if the system oscillation is described by Eq. (A.19), the right-hand side of Eq. (A.20) can be represented as:

$$\sum \Phi = \left(\omega^2 - \omega_0^2\right) a \sin \omega t - 2\delta_0 \psi(a \sin \omega t, a\omega \cos \omega t) a\omega \cos \omega t \tag{A.21}$$

Eq. (A.21) represents a periodic force for the interval Δt. It can be expanded into a Fourier series. The coefficients of this series will be the functions of amplitude and frequency. Then, we have

$$\sum \Phi = f(a, \omega) \sin \omega t + F(a, \omega) \cos \omega t + A_B + \sum \text{Harmonics} \tag{A.22}$$

The constant term A_B, if it exists, indicates a change in bias operating point. If we assume that the oscillating system has a high Q-factor, the amplitude and frequency of the main oscillation harmonic can be found from the equation:

$$\frac{d^2v}{dt^2} + \omega^2 v = f(a, \omega) \sin \omega t + F(a, \omega) \cos \omega t \tag{A.23}$$

It was assumed that the oscillation frequency is constant. In addition, the solution Eq. (A.23) may be arranged to include the sinusoidal component only (which can be obtained by proper choice of oscillation start time). Then one can conclude, using Eq. (A.10) and Eq. (A.11), that:

$$f(a, \omega) = 0 \tag{A.24}$$

and

$$\frac{da}{dt} = \frac{1}{2\omega} F(a, \omega) = 0 \tag{A.25}$$

The last equation describes the amplitude transient for nearly sinusoidal system oscillations. It was obtained by B. van der Pol [1], the Eq. (A.15) (especially the special form considered below) is called van der Pol equation, and the system described by this equation is called van der Pol oscillator. Hence, for calculations of the steady-state amplitude and frequency, one has the system of two nonlinear algebraic equations:

$$f = \begin{cases} f(a, \omega) = 0 \\ F(a, \omega) = 0 \end{cases}$$

That is, the coefficients in Eq. (A.23) are equal to zero.

Reference

1. B. van der Pol, The nonlinear theory of electric oscillations. Proc. Inst. Radio Eng. **22**(9), 1051–1086 (1934)

Index

J. C. F. d. A. Casaleiro et al., *Quadrature RC–Oscillators*, Analog Circuits and Signal Processing, https://doi.org/10.1007/978-3-030-00740-9

MIX
Papier aus verantwortungsvollen Quellen
Paper from responsible sources
FSC® C105338

If you have any concerns about our products,
you can contact us on
ProductSafety@springernature.com

In case Publisher is established outside the EU,
the EU authorized representative is:
Springer Nature Customer Service Center GmbH
Europaplatz 3, 69115 Heidelberg, Germany

Printed by Libri Plureos GmbH
in Hamburg, Germany